热带雨林的奇花 — 寄生花

Himalaya Sapria *(Sapria himalayana)*

—a rare flower found in tropical rain forests

中国热带野生花卉

WILD TROPICAL PLANTS IN CHINA

西双版纳国家级自然保护区管理局
云 南 省 林 业 科 学 院 编
中 国 世 界 语 出 版 社 出版

Compiled by: Xishuangbanna National Natural Reserve Administration,
Yunnan Academy of Forestry Sciences

Published by: China Esperanto Press

First Edition 1998

ISBN 7-5052-0395-9/Q.11

Published by the China Esperanto Press, P.O. Box 77, Beijing, China
Distributed by China International Book Trading Corporation(Guoji Shudian)
35 Chegongzhuang Xilu, Beijing 100044 , China
P.O. Box 399, China
09800

Printed in the People's Republic of China

前　　言

地球南北回归线之间的热带地区，大部分是海洋，有限的陆地除部分岛屿和南美亚马逊平原外，许多地方被沙漠覆盖，绿色植物比较匮乏，而处在北回归线南侧的亚洲部分地区和中国境内，则是另一番景象，那里江河纵横，山峦叠翠，生机盎然，宛如颗颗绿宝石镶嵌在大地上。

中国的热带地区东起东经 123° 附近的台湾省静浦以南，西至东经 85° 的西藏亚东、聂拉木附近。南段纬度在北纬4°（曾母暗沙），其北界蜿蜒于北纬21-24° 之间，基本处在北回归线以南，由于受大气环流以及地形、地貌等诸多因素的影响，在西藏东南部，则升到北纬 28-29° 之间。

中国热带区域的地貌类型错综复杂，有冲积平原、丘陵、山地、高原、石灰岩峰丛、珊瑚岛等，地势从东到西逐渐上升。南海诸岛海拔100米以下，广东东部广布海拔150米以下的丘陵台地，广西多为海拔500-600米的石灰岩构成的峰林，云南东南部热带上限海拔为400米，滇西南达700米，滇西则上升到800米。在焚风(Fotu)效应影响下的金沙江、澜沧江等干热河谷中，纬度更靠北，其海拔上限达1700米。总的来说，中国热带植被基本包括了海南、台湾、广东、广西、云南和西藏等省、区的南部地区。中国陆地的热带雨林区域呈现从东南到西北蜿蜒曲折的斜长带状。

中国热带地区大都濒临海洋。在印度洋西南季风和南太平洋东南季风的影响下，气候湿热，降雨充沛，年平均气温在20-22°，≥10°C的年积温为7500-9000°C以上；年降雨量大多超过1500毫米，部分地区由于受地形影响，可高达5000毫米。这些地带土壤类型多样，植物生长发育良好。除南海诸岛受海浪及台风影响，自然植被发育较少外，其它许多地区如云南南部至东南部、海南岛、广西南部等地，其生物种类极其繁多，生物产量亦最高。这些地区属于世界三大热带雨林之一的东南亚热带雨林的北缘，由湿润雨林、季节性雨林、山地雨林和季雨林四个类型组成，统称为热带北缘雨林。

在中国的热带雨林中，森林生态系统保持最完整、生物资源最丰富、面积最大的原始林区当首推西双版纳。西双版纳土地面积只占国土总面积的1/500，高等植物却占全国总数的1/6，达5000余种，种类之多，居全国之冠，被誉为“植物王国”中的明珠。西双版纳地处横断山脉南部，地势北高南低。北部有哀牢山、无量山为屏障，阻挡了北来的寒潮；南部许多河谷便于西南季风伸入，高温、静风、湿度大。有效的水热配置，使西双版纳成为生物进化的摇篮，为热带花卉的生长发育提供了极好的条件。

和中国其它热带地区一样，西双版纳也属于古热带区。这个地区受第四纪冰川影响较小，古老的植物种类保存较多，在20个古热带的残遗属里，西双版纳就保存有12个属，其中有些还是现代植被组成的重要成份。根据植物属的地理分布性质归化统计，西双版纳有84.4%的属为热带分布型的植物。

热带雨林是地球上最大的生物基因库，也是物种最繁茂、结构最复杂、最生机勃勃的森林类型，在特定的、得天独厚的自然条件下，形成了与其它气候带森林截然不同的一些生物、生态学特征和特性。它们树木高大，群落层次结构复杂，从高达70多米的林中巨人到地上的苔藓、地衣，有5-6层之多，占满了整个空间。它们相互依存又各得其所，充满勃勃生机。发达的板状根支撑着巨大的树体，高高在上，粗壮的气生根，支柱根环绕母树，形成"独木成林"，各种大型木质藤本如手臂、如扁担，最粗的50厘米，最长的达100余米，上下缠绕，左右盘旋，蜿蜒于林间，将林内植物彼此相联;寄生、附生植物密集丛生，组成五彩缤纷的"空中花园";"老茎生花"随处可见，多种榕属植物在枝干上结出了如玛瑙般的串串果实;那因自身繁衍而投靠它木，为生存竞争而致死对方的"绞杀"植物，在雨林中比比皆是;林中长满了大野芋(Alocasia macrorrhiza)、野芭蕉(Musa acuminata)等巨叶植物。这一切，把热带雨林组成了一座琳琅满目、令人眼花缭乱、千变万化的植物迷宫。

热带雨林还是一座大花园，花卉种类繁多，花型奇特，色彩艳丽，气味怪异，五彩缤纷，争艳斗奇。有的花形似蝴蝶，有的似虎须、似羊角、似猫耳、似蜘蛛、似猫须……，和艳丽花朵共生的植物叶片，更是奇形怪状，有的如羊蹄、如鱼尾、如龟背，有的叶还具有滴水尖叶，还有那香飘林间的九里香(Murraya paniculata)和怪味袭人的疣柄魔芋(Amorphophallus virosus)，更有那具红、绿、黄三色的嘉兰(Gloriosa superba)，红、白、黄三色的使君子(Quisgualis indica)，苞片形如花瓣的玉叶金花(Mussa enda spp.)，也有夜间才展露芳容的照夜白(Nyctocalos runfelsiiflora)，有闻歌起舞的风流草(Codariocalyx motorius)，有触动即闭合的含羞草(Mimosa pudica)，有形如舞女在翩翩起舞的舞女姜(Globba recemosa)……。许多落叶乔木的换叶期又逢盛花期:如紫铆(Butea monosperma)、云南石梓(Gmelina arborea)、粉花羊蹄甲(Bauhinia variegata var.Candida)、火筒树(Firmiana colorata)等常常繁花满树，使群落外貌极富生气，华丽异常。热带花卉用它们那千姿百态的花形，五彩斑斓的花色，浓郁宜人的气味招蜂引蝶，繁衍后代。此谢彼盛，四时鲜花不断。

中国热带野生花卉在兰科(Orchidaceae)、姜科(Zingiberaceae)、蝶形花科(Papilionaceae)、天南星科(Araceae)、苦苣苔科(Gesneriaceae)、凤仙花科(Balsaminaceae)、秋海棠科(Begoniaceae)中较为丰富。

人们常把兰科植物称作"热带花王"，因为兰花种类多，数量大，花型奇特，花色美丽。兰花有地生、附生和腐生三大类。生长在热带地区的兰花大部分是附生兰，多生长在树干上或枝丫间。它们利用自己肥厚且带根被的气生根，从空气中吸收水分以及溶于水的无机盐或根系周围的腐殖质和有机物，大树只是作为其繁衍、生长、开花的支撑物。它们常和其他花卉以及附生植物共同组成热带雨林特有的生态现象——"空中花园"，景象十分动人。据专家统计，仅西双版纳就有野生兰花98属351种之多。

热带雨林中各种植物相伴而生。使君子科的千果榄仁(Terminalia myriocarpa)是热带雨林的上层树种。它9-10月开花，花为淡黄色，小巧玲珑，无数小花沿下垂的花序轴密集排列，若干花序簇拥着挂在树端，宛如礼花倒悬。当花凋落后，成熟的红色顶生果序，挂满树冠，有如下垂红缨，非常醒目。

有的藤本如清明花(Beaumontia grandiflora)，也能爬到树冠顶部，而多种榕属植物如木瓜榕(Ficus auriculata)、聚果榕(Ficus racemosa)等，则将它们的花开在茎干上，密密麻麻，十分美丽，蛇菰(Balanophora harlandii)和寄生花(Sapria himalayana)却在树根部寄生。

热带雨林由于森林茂盛，因而林下潮湿、幽暗、阴森，这种遮天蔽日的环境最适宜苔藓、地衣以及蕨类、大叶草本如海芋(Colocasia gigantia)和草花植物如老虎须(Tacca chantrieri)生长。为了争夺阳光，它们互相拥挤、攀附、纠缠，使有限的空间得以充分利用。在这些植物中，草本花卉尤显娇艳多姿，如姜科舞女姜(Globba recemosa)，6-9月开花，花为杏黄色。它的花型侧生退化雄蕊突出花管，并发育成鞭状(又名花丝)，向上反曲。奇怪的是，花丝上端着生出翼状花瓣和象兰花一样的带紫红斑点的唇瓣，花丝的末端着生出鸟喙一样的附属体——花药，整体造型既像舞女抖绸，又像孔雀引颈回眸，实在别致有趣。还有一种百合科的罗甸蜘蛛抱蛋(Aspidistra luodianensis)，其花形硕大，裂片为肉质，花未开时直立，盛开时向外反曲，环抱圆形花筒，颇似爬行的蜘蛛，真是奇特。

热带花卉不仅具有很高的观赏价值，其本身还有众多的经济用途。据初步统计，仅西双版纳自然保护区的热带雨林中就有药用植物139科728种,占中国药用植物总数的14.2%。姜科(中国有19属143种),其中许多种是著名的南药,如姜黄(Curcuma longa)、郁金(Curcuma aromatica)等。另有用材、蔬菜、油料、芳香油、染料、淀粉、树脂、纤维等植物1572种,占高等植物总数的45%,其经济花卉总数多达1500种。可以说热带林中丰富的植物种群是热带花卉用不尽的源泉和种质资源宝库,有着极高的经济开发价值。

热带雨林是地球生物圈的重要组成部分,在维护生态平衡和自然环境方面起着非常重要的作用,也是人类赖以生存的条件。遗憾的是,地球上的热带森林正在遭受着不间断的、不同程度的破坏和毁灭,许多珍稀宝贵的热带花卉也随之消失。据保守的数字统计,每天有 1 万多公顷的热带雨林从地球上消失,几个小时就有一种植物灭绝,取而代之的是次生林,乃至荒漠。中国热带的许多地区由于经济活动的干扰,大面积的雨林被次生类型植物所取代,现在只有在自然保护区和极偏僻的边远山区有小片残林分布。

早在本世纪50年代,自然资源遭破坏的情况即引起了中国政府的重视,拟定了《关于天然森林禁伐区(自然保护区)划定草案》;1978年成立了中国"人与生物圈"国家委员会;1987年5月,国务院环境保护委员会又发布了《中国自然保护纲要》。从1956年在广东鼎湖山建立起中国第一个自然保护区以来,迄今中央和地方政府在全国建立了926个自然保护区,仅云南省在热带地区就建立了20个。云南省西双版纳、南滚河、广东省鼎湖山、海南省尖峰岭、西藏墨脱等自然保护区早已蜚声海内外,并引起有关国际组织的关注和重视。西双版纳和鼎湖山自然保护区由于其丰富的物种资源和独特的生态环境而被联合国教育科学文化组织(UNESCO)列入"世界生物圈保护区网络"。

目前在中国,人们已逐步形成热爱大自然、保护大自然、合理开发大自然的共识,美丽而丰富的热带野生花卉正在日益受到人们的重视,并将在经济发展中起到它不可忽视的作用。

FOREWORD

Most part of the torrid zone between the Tropic of Cancer and the Tropic of Capricorn on the earth is water, and great part of the tropical land areas, except some islands and the Amazon Plains, is covered by desert. Few green plants can be seen there. But some parts of Asia and inside China south of the Tropic of Cancer present another sight—criss-crossing rivers and green mountains.

The torrid zone in China is located basically south of the Tropic of Cancer—from south of Jingpu in Taiwan at 123°E in the east to a spot near Yadong (Chomo) and Nyalam at 85° E in the west, and from Zengmu'ansha Islands at 4°N in the south to an area between 21°and 24°N in the north. Influenced by atmospheric circulation, topography, landforms and many other factors the tropical line may rise to 28°-29°N in southeastern Tibet.

The landforms in the torrid zone in China are complex, with alluvial plains, hilly land, mountain areas, highlands, limestone peaks and coral islands. The terrain rises from east to west. The islands in the South China Sea are below the altitude of 100 meters; tablelands bellow 150 meters are found in large stretches in eastern Guangdong Province; Guangxi Zhuang Autonomous Region has many clusters of limestone peaks of 500 to 600 meters; the highest place in the tropical zone in southeastern Yunnan Province is at the altitude of 400 meters; the highest place in southwestern Yunnan Province is 700 meters above sea level while the altitude in western Yunnan Province rises to as high as 800 meters. Influenced by the Fotu effect the tropical climate may reach 1,700 meters above sea level in the northern section of the hot Jinsha and Lancang river valleys. Generally speaking tropical vegetation in China spreads over the southern part of the provinces of Hainan, Taiwan, Guangdong and Yunnan, and the autonomous regions of Guangxi Zhuang and Tibet. The tropical rain forests in China form a strip tilting from southeast to northwest.

Most part of the tropical zone in China is near the sea. Under the influence of southwesterly monsoon of the Indian Ocean and the southeasterly monsoon of the Pacific Ocean the tropical zone in China has a wet climate with plenty of rainfall. Its annual mean temperature is between 20°and 22°C; the accumulated annual temperature is between 7,500 and 9,000° C; and its annual precipitation is more than 1,500 millimeters and that in some places may be as much as 5,000 millimeters due to local terrain features. Plants grow very well in diverse types of soil in these places. Less vegetation is found in the islands in the South China Sea due to too high tides and too often typhoons. Other places such as the area from southern to southeastern Yunnan Province, Hainan Island and southern Guangxi Zhuang Autonomous Region, boast a great variety of plants, and the biological output there is very high. These places are the northern edge of the Southeastern Asian Tropical Rain Forests, one of the world's three largest tropical rain forest regions in the world, and are divided into four categories of wet rain forests, seasonal rain forests, mountain rain forests and monsoon rain forests, with a general name "tropical northern edge rain forests."

Primeval forests with the best preserved forest ecosystems and forests with the richest biological resources and largest area in China are located in Xishuangbanna, Yunnan Province. With a territory of one 500th of the national total, Xishuangbanna has more than 5,000 species of higher plants, accounting for one sixth of the national total, and for this Xishuangbanna is called a pearl in the "kingdom of plants."

Xishuangbanna lies in southern part of the Hengduan Mountains, rising from south to north. The Ailao and Wuliang mountains in the north prevent cold waves from coming to the south and many river valleys in southern Xishuangbanna lead southwesterly monsoon into the area, causing high temperatures, still winds and high humidity. Effective disposition of water and heat makes Xishuangbanna a cradle of biological evolution and provides excellent conditions for the growth of tropical flowers and plants.

Like other tropical areas in China, Xishuangbanna belongs to the palaeotropical zone. This region bears little glacier impact of the Quaternary Period and has thus preserved a fairly large number of ancient plant species. Of the 20 relic genera of the palaeotropical zone 12 are found in Xishuangbanna and some of them are important elements of the modern vegetation, such as sago cycas *(Cycas spp.)* According to statistics of the plant genera according to geographical distribution, 84.4 percent of the plant species in Xishuangbanna are tropical plants.

Tropical rain forests are the largest biological gene bank on the earth and the most flourishing, complex and energetic type of forests. Their unique, rich natural resources produce some biological and ecological features that are entirely different from forests in other climate zones. Trees there are high and population structure is complex—with as many as five layers from more than 70 meter-high trees to moss and lichen, taking up all the space in the forests. They grow healthily and coexist in harmony. Well-developed plate-like roots support some giant trees which dominate the upper part of the forest; strong and thick air roots and pillar roots hang down from the mother tree and around it to form a "forest"; various vine plants, some 50 centimeters thick and 100 meters long, twine around and climb up trees; parasitic and epiphyte plants form a colorful "air-borne garden;" the sight of flowers blooming on aging stems is common; fig trees bear strings of fruit like rubies on their trunks; the sweet and fragrant fruit of the jackfruit can be as heavy as 10 kilograms; numerous "strangling" plants fight each other for survival; and the wild giant alocasia *(Alocasia macrorrhiza)*, the wild Japanese banana *(Musa acminata)* and the begonia *(Begonia)* with their huge leaves grow in gullies in the forests. Trees and plants of various kinds form a dazzling and ever-changing labyrinth.

Tropical rain forests are a huge garden with many species. Flowers have strange forms, bright colors and peculiar smells. Some flowers look like butterflies, some like tiger's whiskers, some like cat's ear, some like spiders and still some like cat-whiskers. Some leaves resemble sheep hoofs, some fish tail and still some the back of a turtle. Some leaves drip water from their tips. The fragrance of the jasminorange *(Murraya paniculata)* spreads far and wide in the forest, and the strange smell of the giantarum *(Amorphopha virosum)* provokes people's nostril. There are the glorylily *(Gloriosa auperba),* the Rangooncreeper *(Quispulia indica)* in red, white and yellow colors, the Jadeleaf goldflower *(Mussa enda spp.)* with bracts like petals, the broadleaved epiphyllum *(Epiphyllum oxypetalum)* that blooms in early morning, the lighting the night *(Nyctocalos runfelsiiflora)* that blooms only at night; the graceful grass *(Codariocalyx motorium)* that dances when it hears singing, the sensitive plant *(Mimosa pudica)* that closes its leaves when touched, and the dancing-girl ginger *(Globba recemosa)* that looks like a dancing girl. Many deciduous trees bloom and shed their leaves at the same time, such as the Bengal kino *(Butea monosprma),* the Malay bushbeech *(Gmeliina arborea)*, the purple bauhinia *(Bauhinia varlegata)* and the firetube tree *(Firmiana colorata).* Their crowns are covered with blossoms whose strong fragrance attract butterflies. Flowers are found in tropical rain forests in all the four seasons.

Most of tropical plants in China belong to the families of *Orchidaceae*, *Zingiberaceae*, *Papilionaceae*, *Araceae*, *Gesneriaceae*, *Balsaminaceae* and *Begoniaceae*. People often call the orchid "King of Tropical Flowers" for its many varieties, large number and beautiful and peculiar flower forms. Orchids are divided into three major kinds—ground, epiphyte and saprophytic. Most of orchids growing in tropical areas are epiphytes on tree trunks or branches. They absorb moisture, inorganic salts dissolved in water, humus and organic matters through their fat and thick air roots. The tree where they grow is merely a medium for their reproduction, growth and flowering. Orchids and other flowering plants make up an attractive mid-air garden. According to specialists there are 351 kinds in 98 genera in Xishuangbanna.

Various plants coexist in tropical rain forests. The terminalia *(Terminalia myriocarpa)* of the *Quispulia indica* family is a typical higher-layer species in tropical rain forests. Its yellow, exquisite blossoms open in September and October. Numerous tiny flowers hang down the floral axis and a number of inflorescence hang from the tree top like strings of firecrackers. After the flowers wither ripe fruits with a red tip cover the tree crown like red glistening tassels.

The Easter heraldtrumpet *(Beaumontia grandiflora)*, a kind of vine can climb trees; the eared strangler fig *(Ficus auriculata)* and the fruit-gathering fig *(Ficus Racemosa)* put out flowers on their trunks in close arrangement; the snake mushroom *(Balanophora harlandii)* and the parasitic flower *(Sapria himalayana)* grow at the roots of trees.

The ground in dense tropical rain forests is wet and dark, suitable for the growth of moss, lichen, pterdophyte, large-leaved grass such as the elephantsear *(Colocasia gigantia)* and flowering grass such as the tacca *(Tacca chanfrieel)*. In order to compete for sunlight and space, the plants grow closely and entwine with each other. Among them grass flowers are prettier. Dancing-girl ginger *(Globba recemosa)*, for example, blooms from June to September. Its apricot-yellow blossoms grow on one side and its degenerated stamen protrudes from the floral tube like a whip (called floral string) that goes upward and then bends downward. A strange phenomenon is that at the top of the whip-like floral string grows a bird's bill-like appendage—wing-like petals and labellums with purple speckles like those on orchid flowers. The form of the plant looks like a dancing girl playing with ribbons or a peacock looking backward. The common aspidistra *(Aspidistra luodianensi),* a kind of lily, has very large flowers with fleshy petals. Before opening the flower stands erect, but when blooming the petals turn outward and then downward to embrace the circular spline resembling a crawling spider.

Tropical plants are not only decorative but also have high economic value. According to incomplete statistics, 728 kinds of 139 families in the Xishuangbanna Natural Reserve are of medicinal value, accounting for 14.2 percent of medicinal herbs of the whole country. Many ginger varieties (there are 143 kinds of 19 genera in China) are important ingredients in traditional Chinese medicine, such as the common turmeric *(Curcuma longa)* and *(Curcuma aromatica)*. In Xishuangbanna there are 1,572 kinds of timber, vegetable, oil, perfume, dyeing, starch, resin and fiber plants, accounting for 45 percent of the total higher plants. Xishuangbanna also boasts 1,500 commodity flowering plants. The rich plant variety is an inexhaustible source for tropical flowers and a treasurehouse of species. They are of a very high value for commercial development.

Tropical rain forests, an important part of the earth's biosphere, play a very important role in safeguarding ecological balance and natural environment, and provide conditions humankind relies on for its existence. It is a pity tropical rain forests on the earth are being damaged and destroyed uninterruptedly. Many rare tropical flowers have disappeared. According to conservative statistics, in present world about 10,000 hectares of tropical rain forests are eliminated every day and a plant species becomes extinct in several hours. They are replaced by second growth or even by deserts. Many tropical areas in China have been disturbed by human activities and large stretches of rain forests have been replaced by second growth. Today only small stretches of remnant rain forests can be found in natural reserves or extremely remote mountain areas.

As early as in the 1950s, destruction of natural resources had aroused the attention of the Chinese government which formulated the "Draft Plan for the Delimitation of Natural Lumbering Forbidden Area (Natural Reserve)." In 1978 the State "Man and Biosphere" Commission of China was set up. In May 1987 the Environmental Protection Commission under the State Council promulgated the "Program for Natural Protection of China." Since China set her first natural reserve at Dinghushan in Guangdong Province in 1956 the central government and local governments have established more than 926 natural reserves in the country. YunnanProvince alone has 20 natural reserves in its tropical areas. The natural reserves at Xishuangbanna, Nangunhe in Yunnan Province, Dinghushan in Guangdong Province, Jianfengling in Hainan Province and Medog in Tibet Autonomous Region have become well known in the world and attracted attention form relevant international organizations. Those in Xishuangbanna and Dinghushan are on the UNESCO's list of "World Biosphere Protection Network" for their rich species source and unique ecological environment.

The Chinese people are more and more aware of the importance in loving, protecting and rationally developing Great Nature. They pay more and more attention to the beautiful tropical plants and flowers in their large variety, and believe they will play a great role in China's economic development.

西双版纳热带原始森林是中国唯一的,也是地球同纬度地带保存较好的原始热带森林。

Tropical primeval forests in China can be found only in Xishuangbanna in Yunnan Province. They are among the best-preserved tropical primeval forests on the same latitude in the world.

依山傍水，与自然融为一体的傣族村寨。

A village of Dai ethnic minority in Yunnan Province.

云南西双版纳原始热带雨林内，是亚洲象的重要栖息地。

The tropical rain forests in Xishuangbanna are an important habitat of the Asian elephant.

热带雨林中的绞杀植物。

Some “strangler plants” in rain forests.

“绞杀”是热带雨林中的特殊生态现象，反映了热带植物竞争的激烈程度。其产生原因，是种子通过动物传播到某些大树枝杈处而萌发生根，渐成根网，并紧紧包围原附生树干，直到将原树绞死而取而代之。

Seeds are brought by animals and birds to crotches of a big tree where they germinate and take root. Growing up the new plants' roots wrap tightly around the trunk of the big tree and gradually strangle it to death. It is another unique biological phenomenon in tropical rain forests.

板根是热带林中的一个特殊生态景观。图为四数木从树干基部延伸出翼状树根，形如板墙（俗称板根），高达 5 米，延伸到 15 米以外。

Plate roots are a unique biological phenomenon in tropical rain forests. In the picture wing-like roots grow from the base of a trunk like a wall of wooden boards. They are five meters high and 15 meters long.

独树成林是热带雨林的特殊景观。桑科的许多榕属树木，从茎干和枝节上长出不定根和气生根直插地表，从土壤中吸取养分，形成形状各异的支柱根，状如树干。图为云南省盈江县的一株巨榕，有支柱根114根，荫地面积3698平方米，可供千人憩息。

Many adventitious and air roots grow from branches and the trunk of a single banyan tree. They gradually reach the ground and grow into trees. Groves generated from a single tree are a common sight in a tropical rain forest. This giant banyan tree in Yingjiang County, Yunnan Province, has 114 trunks developed from roots. Its crown shades an area of 3,698 square meters under which about 1,000 people can sit.

树包石——热带林奇观。

"Tree embracing a rock", a marvelous sight in tropical rain forests.

扁担藤（Tetrastigma planiculatum）为大型木质藤本植物，是热带雨林的标志特征。图为粗大如臂、穿梭往返的扁担藤。

Shoulder-pole rattan *(Tetrastigma planiculatum)*, a large variety of rattan, is a symbol of tropical rain forests. The vines of the rattan in the picture are as thick as a man's arm.

在湿热的雨林环境中，有些树干上常会有大型附生蕨类生长，如鸟巢蕨（Neottopteris nidus），鹿角蕨（Platycerium wallichii）等。图为多年生蕨类植物王冠蕨（Pseudodrynaria coronans），槲蕨科（Drynariaceae），槲蕨属（Pseudodrynaria）。叶成辐射状，环生于茎干周围，形如王冠而得名。

Some ferns grow on tree trunks, such as the bird's nest fern *(Neottopteris nidus)* and deerhorn fern *(Platycerium wallichii)*. The picture shows the crowan fern *(Pseudodrynaria coronans)* of the *Drynariacceae* family, *Pseudodrynaria* genus. Its leaves radiate from short stems like a crown.

许多乔木树种在茎干和枝干上开花结果，形成热带雨林的又一特有生态现象——茎花现象。

In a rain forest many trees grow flowers and fruit on the trunk or main branches.

鹿角蕨(Platycerinm wallichii),鹿角蕨科(Platyceriaceae),鹿角蕨属(Platycerium),附生观叶植物。因其叶丛生,形如鹿角而得名。它生于云南西部海拔300-700米的林中树上或岩石上。

Deerhorn fern *(Platycerimm wallichii)* of the *Platyceriaceae* family, *Platycerium* genus, is an epiphyte of pteridophyte. Its leaves grow in a cluster resembling a deerhorn, thus its name. It grows on trees or rocks in western Yunnan Province 300 to 700 meters above sea level.

清香木(Pistacia weinmannifolia),漆树科(Anacardiaceae), 黄连木属(Pistacia)。产云南海拔1000－2700米的石灰岩山地或干热河谷中,是中国产木材中最重的乔木,材质优良,根、叶、树皮药用,叶含芳香油。花期4－6月。

Yunnan pistache *(Pistacia weinmannifolia)* of the *Anacardiaceae* family, *Pistacia* genus, is found in limestone mountains or dry gullies 1,000 to 2,700 meters above sea level in Yunnan Province. It produces the heaviest wood in China. Its roots and leaves are of medicinal value, and aromatic essence can be extracted from its leaves. The flowering period is from April to June.

鸭嘴花(Adhatoda vasica),爵床科(Acanthaceae),鸭嘴花属(Adhatota)。产云南,生于海拔850-1800米处。中国仅此一种,全株药用,可驱风镇痛,祛湿消炎。花期4－6月。

Malabarnut *(Adhatoda vasica)* of the *Acanthaceae* family, *Adhatota* genus, grows in places 850 to 1,800 meters above sea level in Yunnan Province. China has only one variety. The whole plant can be used as medicine, effective to release pain and inflammation. It blooms from April to June.

假杜鹃(Barleria cristata)，爵床科(Acanthaceae)，假杜鹃属(Barleria)。产中国南部和西南部，生于海拔900-1700米的干旱山坡或灌丛中。草本，全株治蛇伤、关节痛，亦为绿化花木。花期2-6月。

Barleria *(Barleria cristata)* of the *Acanthaceae* family, *Barleria* genus, is found in southern and southwestern China on dry mountain slopes or on bushes 900 to 1,700 meters above sea level. The herb is effective in treating snake poison and arthritis. Its flowering period lasts from February to June.

喜花草(Eranthemum pulchllum)，爵床科(Acanthaceae)，喜花草属(Eranthemum)。产云南西双版纳沟谷密林下。花期2－3月。

Happy flower *(Eranthemum pulchllum)* of the *Acanthaceae* family, *Eranthemum* genus, grows in dense forests in deep gullies in Xishuangbannan of Yunnan Province. Its flowering period lasts from February to March.

弯花焰爵床（Phlogacanthus curiflora），爵床科（Acanthaceae），火焰花属（Phlogacanthus）。产云南南部，生于海拔500-900米的密林下及沟边潮湿的地方。花期3-4月，花大而密集，为极好的观赏植物。

Firework acanthus *(Phlogacanthus curiflora)* of the *Acanthaceae* family, *Phlogacanthus* genus, grows in dense forests or wet spots in southern Yunnan Province 500 to 900 meters above sea level. Its big, closely clustered flowers bloom in March and April. It is a very attractive flowering plant.

大花山牵牛（Thunbergia grandiflora），爵床科（Acanthaceae），山牵牛属（Thunbergia）。产云南、广东、广西，生于海拔600-900米的林缘或灌丛中。花大且美丽，是极好的观赏花卉，花期2-4月。根药用。

Bengal clockvine *(Thunbergia grandiflora)* of the *Acanthaceae* family, *Thunbergia* genus, is found on the edge of a forest or in bushes 600 to 900 meters above sea level in the provinces of Yunnan and Guangdong and Guangxi Zhuang Autonomous Region. Its big and beautiful flowers bloom from February to April. Its root is of medicinal value.

柬埔寨龙血树(Dracaena cambodiana),龙舌兰科(Agavaceae),龙血树属(Dracaena)。产海南、云南南部,生于海拔800－1300米的石灰岩山上石缝中,老树木质中可提取药用血竭。花期10－12月。

Cambodian Dragon Dracaena *(Dracaena cambodiana)* of the *Agavaceae* family, *Dracaena* genus, grows in crevices in limestone mountains 800 to 1,300 meters above sea level in Hainan Province and southern Yunnan Province. The essence extracted from old trees is used as an ingredient of Chinese medicine. Its flowering period lasts from October to December.

云南蕊木(Kopsia officinalis),夹竹桃科(Apocynaceae),蕊木属(Kopsia)。特产云南西双版纳,生于海拔500-800米的山地疏林中。树皮煎水可治水肿;果实和叶有消炎止痛、舒筋活络之效,又可治咽喉炎、扁桃腺炎、风湿骨痛、四肢麻木等症。花期4-9月。

Yunnan kopsia *(Kopsia officicinalis)* of the *Apocynaceae* family, *Kopsia* genus, is a special local plant of Xishuangbanna in Yunnan Province. It grows in sparse woods 500 to 800 meters above sea level. Its bark is boiled to treat dropsy, and its fruit and leaves are effective in reducing inflammation and pain and in treating pharyngitis, tonsillitis, arthritis and limb numbness. It blooms from April to September.

清明花（Beaumontia grandiflora），夹竹桃科（Apocynaceae），清明花属（Beaumontia）。产云南西双版纳，生于海拔720-800米的中山或沟谷密林中。其根和叶可治风湿和瘀肿。花期3－4月。

Easter heraldtrumpet *(Beaumontia grandiflora)* of the *Apocynaceae* family, *Beaumontia* genus, grows in mountains or deep gullies in dense forests 720-800 meters above sea level in Xishuangbanna, Yunnan Province. It blooms in March and April, and its roots and leaves can be used as medicine to treat rheumatism and extravasated blood.

毛 车 藤 (Amalocalyx yunnanensis), 夹 竹 桃 科 (Apocynaceae), 毛 车 藤 属 (Amolocalyx)。 产 云 南，生于海拔700-1100米的坡地、林缘灌丛中。花期5-6月。

Maocheteng rattan (*Amalocalyx yunnanensis*) of the *Apocynaceae* family, *Amolocalyx* genus, is found on slopes and bushes on the edge of forests in Yunnan Province, 700 to 1,100 meters above sea level. It blooms in May and June.

海芋(Alocasia macrorrhiza),天南星科(Araceae),海芋属(Alocasia)。产福建、台湾、广东、广西、贵州、云南的热带至亚热带地区,生于海拔500-1700米的热带雨林中。根茎药用,对腹痛、霍乱、疝气等有良效。本品有毒,须久煎并换水2-3次后方能服用。花期7-8月。

Giant alocasia *(Alocasia macrorrhiza)* of the *Araceae* family, *Alocasia* genus, grows in tropical and subtropical rain forests 500 to 1,700 meters above sea level in the provinces of Fujian, Taiwan, Guangdong, Guizhou and Yunnan and Guangxi Zhuang Autonomous Region. Its root is of medicinal value, good for treating stomach ach, cholera and hernia. But the poisonous roots must be boiled in water that is replaced two to three times before it can be taken orally.The plant blooms in July and August.

疣柄魔芋(Amorphophallus virosus),天南星科(Araceae),魔芋属(Amorphophallus)。产云南、广东、广西,生于海拔180-700米的河边草地、灌丛或疏林下。花大,美丽,可供观赏。全株用火灰煮后作猪饲料,催膘甚快。花期4-5月。

Wart-handle giantarum *(Amorphophallus virosum)* of the *Araceae* family, *Amorphophallus* genus, grows among grass, bushes or sparse woods near a river 180 to 700 meters above sea level in the provinces of Yunnan and Guangdong and Guangxi Zhuang Autonomous Region. Its flowers are big and beautiful. The whole plant can be boiled and fed to pigs for fattening. It blooms in April and May.

泉 七 (Steudnera colocasiaefolia),天南星科(Araceae),泉七属(Steudnera)。产广西、云南,生于海拔500 - 900米的沟谷林中溪边。花期4 - 5月。根茎药用,但有毒。

Quanqi *(Steudnera colocasiaefolia)* of the *Araceae* family, *Steudnera* genus, grows in tree groves or by streams in gullies 500 to 900 meters above sea level in Guangxi Zhuang Autonomous Region and Yunnan Province. It blooms in April and May. Its poisonous root is of medicinal value.

大叶崖角藤（Rhaphidophora megaphylla），天南星科（Araceae），崖角藤属（Rhaphidophora）。云南特产，在海拔800－1300米热带密林中的大树上和岩石上附生。叶大、革质，茎粗壮，花为佛焰苞，黄白色，极美。全株入药，内治脾肿、外敷治跌打损伤、骨折、风湿腰痛及火、烫伤。花期4－6月。

Big-leaf rock edge rattan *(Rhaphidophora megaphylla)* of the *Araceae* family, *Rhaphidophora* genus, is a special plant of Yunnan Province. An epiphyte on trees and rocks, it has large leaves, hard skin and thick vines. The yellow or white flowers are very beautiful. Every part of the plant can be used as medicine for treating swollen spleen, injuries, broken bones, backache due to arthritis, and burns. Its flowering period is from April to June.

犁头尖(Typhonium diversfolium),天南星科(Araceae),犁头尖属(Typhonium)。产于云南南部，生于海拔500-900米的林缘旷地或田边地角。全草及块茎药用。花期4-5月。

Ploughshare Tip *(Typhonium diversfolium)* of the *Araceae* family, *Typhonium* genus, grows in open lots by the edge of forests or fields 500-900 meters above sea level in southern Yunnan Province. The whole plant with stem tuber can be used as medicine. It blooms in April and May.

南蛇棒(Amorphophallus dunnii),天南星科(Araceae),魔芋属(Amorphophallus)。产广东及沿海岛屿、香港、广西、云南东南部,生于海拔220-800米的林下荫湿处。花期3 - 4月。

Nanshepang giantarum *(Amorphophallus dunnii)* of the *Araceae* family, *Amorphophallus* genus, grows in wet spots in forests 220 to 800 meters above sea level on coastal islands of Guangdong Province, Hong Kong, Guangxi Zhuang Autonomous Region and southeastern Yunnan Province.It blooms in March and April.

滇南魔芋(Amorphallus yunnanensis),天南星科(Araceae),魔芋属(Amorphallus)。产云南、贵州,生于海拔200-2000米的山坡密林中。花期4-5月。

Southern Yunnan giantarum *(Amorphallus yunnanensis)* of the *Araceae* family, *Amorphallus* genus, grows in dense forests on mountain slopes 200 to 2,000 meters above sea level in Yunnan and Guizhou provinces. It blooms in April and May.

广防已（Aristolochia fangchii），马兜铃科（Aristolochiaceae），马兜铃属（Aristolochia）。产云南，生于海拔700-900米的山坡疏林下或林缘灌丛中。根药用，治胃炎、腹痛、消化不良。花期10-11月。

Guangdong root of fangji *(Aristolochia fangchii)* of the *Aristolochiaceae* family, *Aristolochia* genus, grows in sparse woods or bushes on the edge of forests 700 to 900 meters above sea level in Yunnan Province.Its root can be used to treat gastritis,stomach ache and indigestion. It blooms in October and November.

广西马兜铃（Aristolochia kwangsiensis），马兜铃科（Aristolochiaceae），马兜铃属（Aristololochia）。特产广西西南部，生于海拔500－800米的石灰岩山地常绿阔叶林下。块根入药，有小毒，治喉痛、胃痛、肚痛、蛇伤。花期9－10月。

Guangxidutchmanspipe(*Aristolochia kwangsiensis*) of the *Aristolochiaceae* family, *Aristolochia* genns, is found only in southwestern Guangxi Zhuang Autonomous Region in evergreen broadleaf woods 500 to 800 meters above sea level. Its root, slightly poisonous, is used as medicine to treat throat sore, stomach ache and snake bites.It blooms in September and October.

马利筋(Asclepias curassavica),萝藦科(Asclepiadaceae),马利筋属 (Asclepias)产中国热带地区,生于海拔500-1100米的河边村旁和路边灌丛中,全株药用,有毒。花期6-7月。

Bloodflower milkweed *(Asclepias curassavica)* of the *Asclepiadaceae* family, *Asclepias* genus, is a tropical plant found in bushes by the river or near villages 500-1,100 meters above sea level. The whole plant is poisonous but can be used as medicine. It blooms in June and July.

牛角瓜 (Calotropis gigantea), 萝藦科(Asclepiadaceae),牛角瓜属(Calotropis)。产云南、四川、广西、广东,生于低海拔向阳山坡旷地上。花期10 - 12月。

Akund calotrope *(Calotropis gigantea)* of the *Asclepiadaceae* family, *Calotropis* genus, grows in sunny mountain slopes of low altitudes in the provinces of Yunnan, Sichuan and Guangdong, and Guangxi Zhuang Autonomous Region. Its flowering period lasts from October to December.

凸脉球兰 (Hoya nervosa)，萝藦科 (Asclepiadaceae)，球兰属 (Hoya)。特产云南南部海拔600米左右的山地密林中，附生于树上。花期5-7月。

Common waxplant *(Hoya nervosa)* of the *Asclepiadaceae* family, *Hoya* genus, can be found only in southern Yunnan Province in dense forests on mountains at an altitude of about 600 meters. The tree epiphyte blooms from May to July.

通关散 (Marsdenia tenacissima)，萝藦科 (Asclepiadaceae)，牛奶菜属 (Marsdenia)。产云南南部、西部和广西，生于海拔500-1200米山坡疏林或石灰岩山林下。藤茎供药用。花期5-6月。

Condorvine *(Marsdenia tenacissima)* of the *Asclepiadaceae* family, *Marsdenia* genus, grows in sparse woods or in forests on limestone mountains 500-1,200 meters above sea level in southern and western Yunnan Province and Guangxi Zhuang Autonomous Region. Its vine can be used as medicine. It blooms in May and June.

蛇菰科

BALANOPHORACEAE

蛇菰 (Balanophora harlandii)，蛇菰科 (Balanophoraceae)，蛇 菰 属 (Balanophora)。 产云南、广东、广西、四川、贵州等省区，生于海拔800-2000米的林下潮湿环境中。花期9-12月。全株药用，煎服可解毒。

Balano *(Balanophora harlandii)* of the *Balanophoraceae* family, *Balanophora* genus, grows in the provinces of Yunnan, Guangdong, Sichuan and Guizhou and Guangxi Zhuang Autonomous Region in wet spots under trees 800 to 2,000 meters above Sea level. It blooms from September to December. The whole plant can be boiled and drunk to treat poisoning.

大苞凤仙花（Impatiens balansae），凤仙花科（Balsaminaceae），凤仙花属（Impatiens）。产云南南部，生于海拔600-1200米林下沟边草丛中或沙地上。花期5-6月。

Large-bud balsam (*Impatiens balansae*) of the *Balsaminaceae* family, *Impatiens* genus, is found in southern Yunnan Province under trees, in grass and on sandy land 600 to 1,200 meters above sea level. It blooms in May and June.

龙州凤仙花（Impatiens morsei），凤仙花科（Balsaminaceae），凤仙花属（Impatiens）。产于广西龙州，生于海拔600-850米的石灰岩山常绿阔叶林下阴湿处。花期6－7月。

Longzhou basam (*Impatiens morsei*) of the *Balsaminaceae* family, *Impatiens* genus, grows in Longzhou, Guangxi Zhuang Autonomous Region in wet spots under evergreen broadleaf trees on limestone mountains 600 to 850 meters above sea level. It blooms in June and July.

歪叶秋海棠(Begonia angustinei),秋海棠科(Begoniaceae),秋海棠属(Begonia)。产云南南部,生于海拔600-900米的林下水边潮湿的环境中。花期10-11月。

Tilted-leaf begonia *(Begonia angustinei)* of the *Begoniaceae* family, *Begonia* genus, grows in wet areas under trees or by water 600 to 900 meters above sea level in southern Yunnan Province. It blooms in October and November.

厚叶秋海棠(Begonia drgadis),秋海棠科(Begoniaceae),秋海棠属(Begonia)。产云南西双版纳海拔600-1400米的沟谷林中。花期3-4月。

Thick-leaf begonia *(Begonia dragadis)* of the *Begoniaceae* family, *Begonia* genus, grows in woods in gullies 600 to 1,400 meters above sea level in Xishuangbanna, Yunnan Province. It blooms in March and April.

掌叶秋海棠(Begonia hemsleyana),秋海棠科(Begoniaceae),秋海棠属(Begonia)。产云南、广西,生于悬崖旁的溪边和林下半阴处的腐叶土中。多年生草本,属球根类秋海棠,叶多有斑点,是庭园观赏植物。花期4－6月。

Palm-leaf begonia *(Begonia hemsleyana)* of the *Begoniaceae* family, *Begonia* genus, grows by streams or in semi-shade under trees near cliffs in Yunnan Province and Guangxi Zhuang Autonomous Region. The perennial herb with specked leaves, an ideal decorative courtyard plant, blooms from April to June.

翅叶木（Pauldopia ghorta），紫葳科（Bignoniaceae），翅叶木属（Pauldopia）。云南特产，生于海拔1350-1750米的山坡密林中。花期5-6月。

Wingleaf wood *(Pauldopia ghorta)* of the *Bignoniaceae* family, *Pauldopia* genus, is a special local plant of Yunnan Province found in dense forests on mountain slopes 1,350 to 1,750 meters above sea level. It blooms in May and June.

火烧花（Mayodendron igneum），紫葳科（Bignoniaceae），火烧花属（Mayodendron）。产云南、广东、广西、台湾，生于海拔(200-)700-1500米的密林中或比较湿润的河谷低地。茎花植物，开花时满树皆红，故名火烧花。花期3-4月，可作庭园观赏及行道树，花可食用。

Flaming Flower *(Mayodendron igneum)* of the *Bignoniaceae* family, *Mayodendron* genus, grows in dense forests or wet gullies 700 to 1,500 meters above sea level in the provinces of Yunnan, Guangdong and Taiwan and Guangxi Zhuang Autonomous Region. Its flowers on the stems are red and edible. It blooms in March and April. The pretty plant is often used to decorate courtyards and streets.

照夜白（Nyctocalos runfelsiiflora），紫葳科（Bignoniaceae），照夜白属（Nyctocalos）。云南特产，生于海拔160-670(-1500)米干热河谷或河边林中，攀援于树上。夜晚开白色花朵，白天闭合，故而得名。花期3-4月。

Lighting the night *(Nyctocalos runfelsiiflora)* of the *Bignoniaceae* family, *Nyctocalos* genus, is special to Yunnan Province. The vine plant climbs trees in dry and hot gullies or in woods by a river 160 to 670 meters above sea level. Its white blossoms open during the day and close at night in March and April.

岩梧桐（Weightia speciosissima），紫葳科（Bignoniaceae），岩梧桐属（Weightia）。产云南南部至东南部，生于海拔1250-2500米的山坡及疏林中。花期10-12月，盛花时，花满枝头，极为壮观，是庭园绿化的良好树种。根药用，可治跌打骨折、风湿关节炎、四肢麻木、胃痛、月经不调、宫寒不孕、体虚等症。

Chinese parasol *(Weightia speciosissima)* of the *Bignoniaceae* family, *Weightia* genus, is found from southern to eastern Yunnan Province in sparse woods or mountain slopes 1,250 to 2,500 meters above sea level. It blooms from October to December. At the blooming height the crown is a thick layer of flowers, an ideal tree to decorate courtyards. Its root can be used as medicine to treat injuries, broken bones, arthritis, limb numbness, stomach ache, menoxenia, sterility and physical weakness.

木棉 (Bombax ceiba)，木棉科 (Bombacaceae)，木棉属 (Bombax)。主产西南和华南，生于干热河谷和河堤两岸。花期2-3月，花大色红，萼肉质，除观赏外，种子的棉毛可做填充物。

Bombax *(Bombax ceiba)* of the *Bombacaceae* family, *Bombax* genus, grows mainly in dry, hot valleys or on river banks in Southwest and South China. Its large, red flowers open in February and March and the cotton on its seeds can be used as a filling material.

荚迷叶山柑（Capparis viburnifolia），山柑科（Capparidaceae），山柑属（Capparis）。产云南西双版纳、思茅，生于海拔1300米以下较为干燥的地方。花期2－3月。

Peashell caper *(Capparis viburnifolia)* of the *Capparidaceae* family, *Capparis* genus, grows in dry land lower than 1,300 meters of altitude in Xishuangbanna and Simao, Yunnan Province. It blooms in February and March.

蜘蛛花(Silvanthus bracteatus)，四角果科(Carlemanniaceae)，蜘蛛花属(Silvanthus)。产云南南部，生于海拔600-800米雨林下的沟边潮湿的地方，3-4月开花，因萼片形如蜘蛛而得名。

Spider flower *(Silvanthus bracteatus)* of the *Carlemanniaceae* family, *Silvanthus* genus, grows by ditches in rain forests 600 to 800 meters above sea level in Yunnan Province. It blooms in March and April. Its name is derived from the spider-like formation of petals.

使君子科

COMBRETACEAE

使君子(Quisgualis indica)，使君子科(Combretaceae)，使君子属(Quisqualis)。产中国热带地区，生于海拔500-900米的林缘灌丛，根、叶、种子药用;花能变色，供观赏。花期5-6月。

Rongooncreeper *(Quisqualis indica)* of the *Combretaceae* family, *Quisqualis* genus, is a Chinese tropical plant found in bushes on the edge of forests 500-900 meters above sea level. Its roots, leaves and seeds can be used as medicine. Its flowers change color. The beautiful decorative plant blooms in May and June.

千果榄仁(Terminalia myriocarpa)，使君子科(Combretaceae)，榄仁树属(Termialia)。产云南、广西、西藏东南部，生于海拔600-1,500(-2500)米的沟谷热带雨林中。花期9-10月。用材树种。

Bayberry waxmyrtlefruit terminalia *(Terminalia myriocarpa)*, which belongs to the *Combretaceae* family and the *Terminalia* genus, grows in the tropical rain forests in gullies and valleys 600-1,500 meters above sea level in Yunnan, Guangxi and Southeast Tibet. Its flowers come into bloom in September and October every year. It is a source of timber.

七爪龙（Ipomoea digitata），旋花科（Convolvulaceae），番薯属（Ipomoea）。产云南、广东、广西、海南、台湾，生于海拔200-700米的山地林缘和平坝灌丛中。花期7-8月。块根药用，磨碎用于消肿及腹泻，但有毒。

Fingerleaf morningglory *(Ipomoea digitata)* of the *Convolvulaceae* family, *Ipomoea* genus, is found on the edge of forests or in bushes on plains 200 to 700 meters above sea level in the provinces of Yunnan, Guangdong, Hainan and Taiwan. It blooms in July and August. Its root, slightly poisonous, can be ground into powder to treat swelling or diarrhoea.

滇南千里光(Senecio hoi)，菊科(Compositae)，千里光属(Senecio)。产云南南部至东南部，生于海拔1500米以下疏林及灌丛中。花期8－9月。

South Yunnan Senecio *(Senecio hoi)* of the *Compositae* family, *Senecio* genus, grows in sparse woods and bushes below 1,500 meters of altitude from southern to southeastern Yunnan Province. It blooms in August and September.

攀枝花苏铁 (Cycas panzhihuaensis)，苏铁科 (Cycadaceae)，苏铁属 (Cycas)。产四川省攀枝花市、宁南县、德昌县、盐源县以及云南的禄劝县、元谋县和华坪县，生于海拔 1100-1700 米的石灰岩山坡上。花期 4-5 月。图为雄花，花呈圆柱纺锤状，上有密生黄色绒毛，花体硕大、雄壮。

Panzhihua cycas *(Cycas panzhihuaensis)* of the *Cycadaceae* family, *Cycas* genus, grows on limbestone mountain slopes 1,100 to 1,700 meters above sea level in Panzhihua City, Ningnan County, Dechang County and Yanyuan County in Sichuan Province, and Luquan County, Yuanmuo County and Huaping County in Yunnan Province. It blooms in April and May. The staminate flower in the picture is large and looks like a spindle with close fine hair on it.

叉叶苏铁(Cycas micholitzii),苏铁科(Cycadaceae)。苏铁属(Cycas),产广西龙州和云南河口地区,常生于海拔150-300米石灰岩山地疏林或草地灌丛中。花期4 - 5月。

Micholitz *(Cycas micholitzii)* of the *Cycadaceae* family grows in sparse woods or bushes on limestone mountains 150 to 300 meters above sea level in Longzhou, Guangxi Zhuang Autonomous Region, and Hekou, Yunnan Province. It blooms in April and May .

篦齿苏铁(Cycas pectinata)，苏铁科(Cycadceae)。苏铁属(Cycas)，产云南南部红河、思茅和西双版纳，生于海拔1500米以下的常绿阔叶林中，常隔1-3年开花，花期6-7月。图为雌花，花呈扁平球状，密生诸多掌状鳞片，颇富观赏价值。

Nepal cycas *(Cycas pectinata)* of the *Cycadceae* family, *Cycas* genus, grows in evergreen broadleaf forests lower than 1,500 meters of altitude in Honghe, Simao and Xishuangbanna of southern Yunnan Province. It blooms in June and July. Its male flower shown in the picture has flat, palm-like petals.

寄生花(Sapria himalayana),大花草科(Cytinaceae),寄生花属(Sapria)。中国仅此一种,产云南南部和西藏东南部,草本植物,多寄生在海拔700-900米密林中的扁担藤上。花期11-12月。

Himalaya sapria *(Sapria himalayana)* of the *Cytinanceae* family, *Sapria* genus, can be found in southern Yunnan Province and southeastern Tibet Autonomous Region. The herb lives on shoulder-pole vine in dense forests 700 to 900 meters above sea level. It blooms in November and December.

小叶云实（胀荚云实）(Caesalpinia millettii)，苏木科（Caesalpiniaceae），云实属(Caesalpinia)。产云南南部，生于海拔600-700米的林缘灌丛中。花期9-10月。

Small-leaf Brazilwood *(Caesalpinia millettii)* of the *Caesalpiniaceae* family, *Caesalpinia* genus, grows in bushes on the edge of woods, 600 to 700 meters above sea level in southern Yunan Province. It blooms in September and October.

粉花羊蹄甲(Bauhinia variegata Var.candida)，苏木科(Caesalpiniaceae)，羊蹄甲属(Bauhinia)。产云南、广东、广西，生于海拔600-900米的季雨林中。花、嫩荚可作蔬菜。3-4月为盛花期，开花时，遍山皆白。

Buddhist Bauhinia *(Bauhinia variegata var candida)* of the *Caesalpiniaceae* family, *Bauhinia* genus, grows in monsoon rain forests 600 to 900 meters above sea level in Yunnan and Guangdong provinces and Guangxi Zhuang Autonomous Region. Its flowers and tender beans are edible. In its blooming season in March and April, mountainsides are covered with its white flowers.

无忧花(Saraca dives)，苏木科(Caesalpiniaceae)，无忧花属(Saraca)。产云南、华南，生于海拔320－580米的密林中。花状密集，排列为半球形，是极好的庭园绿化植物。花期4-5月。

Common Saraca *(Saraca dives)* of the *Caesalpiniaceae* family, *Saraca* genus, grows in dense forests 320 to 580 meters above sea level in Yunnan Province and South China. Its flowers are in close clusters arranged in a semi-circle and bloom in April and May.

龙脑香科

DIPTEROCARPACEAE

云南龙脑香(Dipterocarpus retusus)，龙脑香科(Dipterocarpaceae)，龙脑香属(Dipterocarpus)。产云南东南部及西部，生于海拔200-900米的热带密林中，为珍贵用材树种。花期5-6月，可供观赏。

Common borneocamphor *(Dipterocarpus retusus)* of the *Dipterocarpaceae* family, *Dipterocarpus* genus, grows in dense rain forests 200 to 900 meters above sea level in southeastern and western Yunnan Province. It is a good timber tree of high value. Its flowering period is in May and June.

望天树(Parashorea chinensis)，龙脑香科(Dipterocarpaceae)，柄安属(Parashorea)。中国特有树种，仅产云南省南部以及广西南部，生于海拔350－1100米河谷及沟谷山麓湿润处，树高可达80米，胸径1米以上，树干通直圆满，木材经济价值较高。本种对中国热带雨林及植物区系研究具有重要科学价值。花期4-5月。

Wangtian tree *(Parashorea chinensis)* of the *Dipterocarpaceae* family, *Parashorea* genus, is a special tree species of China found in southern Yunnan Province and southern Guangxi Zhuang Autonomous Region in river valleys and mountain gullies 350 to 1,100 meters above sea level. It can grow as high as 80 meters with a diameter of more than one meter and blooms in April and May. The trunk is straight and round. The wood is high quality timber. This tree is of special value for the study on tropical rain forests and plant system in China.It blooms in April and May.

苏铁蕨(Brainia insignis),乌毛蕨科(Dryopteridaceae),苏铁蕨属(Brainia)。产云南、广东、广西、香港及台湾,生于海拔900-1300米的季风常绿阔叶林下。根茎药用。

Sago cycas fern *(Brainia insignis)* of the *Dryopteridanceae* family, *Brainia* genus, grows in evergreen broadleaf monsoon rain forests 900 to 1,300 meters above sea level in Yunnan Province, Guangdong Province, Guangxi Zhuang Autonomous Region, Hong Kong and Taiwan Province. Its root is of medicinal value.

水石榕(Elaeocarpus hainanensis)，杜英科(Elaeocarpaceae)，杜英属(Elaeocarpus)。产海南，生于海拔200-600米的水边。花下垂，可作庭院观赏。花期4-5月。

Waterstone banyan *(Elaeocarpus hainanensis)* of the *Elaeocarpaceae* family, *Elaeocarpus* genus, grows by water 200 to 600 meters above sea level in Hainan Province. Its flowers bend toward the ground, ideal for courtyard decoration. It blooms in April and May.

锥花三宝木(Trigonostemon thyrsoideum),大戟科(Euphorbiaceae),三宝木属(Trigonostemon)。产云南南部及西南部,生于海拔400-1000米的密林下及沟边潮湿的环境中。花期4-5月。

Taper-flower three-treasure wood *(Trigonostemon thyrsoideum)* of the *Euphorbiaceae* family, *Trigonostemon* genus, grows in dense forests or in wet land by ditches 400-1,000 meters above sea level in southern and southwestern Yunnan Province. It blooms in April and May.

矮芒毛苣苔(Aeschynanthus humilis)，苦苣苔科(Gesneriaceae)，芒毛苣苔属(Aeschynanthus)。云南特产，生于海拔1200-2100米的林中树上。花色极为美丽，花期10-12月。

Shortprickled basketvine *(Aeschynanthus humilis)* of the *Gesneriaceae* family, *Aeschynanthus* genus, is special to Yunnan Province. It grows in woods 1,200 to 2,100 meters above sea level. Its beautiful flowers bloom from October to December.

大花芒毛苣苔（Aeschynanthes mimetes），苦苣苔科（Gesneriaceae），芒毛苣苔属（Aeschynanthes）。产西藏东南部和云南西部及南部，生于海拔1000-1900米的林中树上。花期6-9月。

Lobecup basketvine *(Aeschynanthes mimetes)* of the *Gesneriaceae* family, *Aeschynanthus* genus, grows on trees in woods 1,000 to 1,900 meters above sea level in southeastern Tibet Autonomous Region and western and southern Yunnan Province. Its flowering period lasts from June to September.

美丽唇柱苣苔(Chirita speciosa)，苦苣苔科(Gesneriaceae)，唇柱苣苔属(Chirita)。产云南南部、中部1450米左右的山间林下。花期7-9月。

Beautiful-lip basketvine *(Chirita speciosa)* of the *Gesneriaceae* family, *Chirita* genus, grows in woods on mountains 1,450 meters above sea level in southern and central Yunnan Province. Its flowering period lasts from July to September.

髯丝蛛毛苣苔(Paraboea rutescens),苦苣苔科(Gesneriaceae),线柱苣苔属(Rhynchotechum)。产云南南部400米以下山岩沟边。花期6－9月。

Thread-pillar paraboea *(Paraboea rutescens)* of the *Gesneriaceae* family, *Rhynchotechum* genus, grows under mountain rocks or by ditches lower than 400 meters in southern Yunnan Province. Its floweirng period lasts from June to September.

朱红苣苔(Calcareoboca coccinea),苦苣苔科(Gesneriaceae),朱红苣苔属(Calcareoboca)。特产云南东南部和广西西部,生于海拔1000-1460米的石灰岩山常绿阔叶林下石上。花期5-7月。

Scarlet basketvine *(Calcareoboca coccinea)* of the *Gesneriaceae* family, *Calcareoboca* genus, is a special local plant of southeastern Yunnan Province and western Guangxi Zhuang Autonomous Region. It grows in evergreen broadleaf forests on limbestone mountains 1,000 to 1,460 meters above sea level.Its flowering period lasts from May to July.

长圆吊石苣苔(Lysionotus oblongifolius),苦苣苔科(Gesneriaceae),吊石苣苔属(Lysionotus)。特产广西西部,生于石灰岩山常绿阔叶林下。花冠紫红色,艳丽。花期6-8月。

Round, hanging-rock basketvine *(Lysionotus oblongifolius)* of the *Gesneriaceae* family, *Lysionotus* genus, is special to western Guangxi Zhuang Autonomous Region. It grows in evergreen broadleaf woods on limestone mountains. Its flowers are bright purplish red.It blooms from June to August.

铁力木(Mesua ferrea)，藤黄科(Guttiferae)，铁力木属(Mesua)。产云南南部和西南部， 生于海拔800-900米的林中。新叶红色，花瓣白色，雄蕊黄色，可作庭园观赏。云南佛寺周围多有种植。花期4-5月。木材坚硬，为名贵用材树种。

Ferreous mesua *(Mesua ferrea)* of the *Guttiferae* family, *Mesua* genus, grows in woods 800 to 900 meters above sea level in southern and southwestern Yunnan Province. Its new leaves are red. Its flowers have white petals and yellow stamens. A good courtyard plant it is often seen near Buddhist temples in Yunnan. It blooms in April and May. Its wood is very hard.

大叶仙茅（Curculigo capitullat），仙茅科(Hypoxidaceae)，仙茅属(Curculigo)。分布于中国西南至东南热带地区，生于海拔700-1000米沟谷密林下。多年生草本，花期10-11月。叶性韧，可缚物，根茎药用。

Largeleaf Curculigo *(Curculigo capitullat)* of the *Hypoxidaceae* family, *Curculigo* genus, is distributed in the tropical zone from Southwest to Southeast China. It grows in woods in deep valleys 700 to 1,000 meters above sea level. The perennial herb blooms in October and November. Its leaves are very durable often used as binding rope. Its root is of medicinal value.

常山（Dichroa febrifuga），八仙花科（Hydrangiaceae），常山属（Dichroa）。产云南南部及西部，生于海拔900-1300米沟谷密林下或水边灌丛中。花期6月。

Antipyretic dichroa *(Dichroa febrifuga)* of the *Hydrangiaceae* family, *Dichroa* genus, grows in dense forests or by waterside bushes 900-1,300 meters above sea level in southern and western Yunnan Province. It blooms in June.

猫须草 (Clerodendranthus spicatus)，唇形科(Labiatae)，肾茶属(Clerodendranthus)。云南、广西、广东、海南、台湾均有分布，生于海拔950-1050米的林中空旷草地及林缘。花期9-10月。为极好的利尿药，用于急性肾炎，对膀胱炎尤为有效；也可用于治疗尿道结石、风湿性关节炎。因其雄蕊酷似猫须而得名。

Cat-whisker grass *(Clerodendranthus spicatus)* of the *Labiatae* family, *Clerodendranthus* genus, is distributed in the provinces of Yunnan, Guangdong, Hainan and Taiwan and Guangxi Zhuang Autonomous Region. It grows in open meadows or on the edge of forests 950 to 1,050 meters above sea level. It blooms in September and October. The plant is of medicinal value, good for treating urinating difficulty, acute urethra stone and rheumarthritis. Its name derives from the shape of its stamens.

光泽锥花（Gomphostema lucidum），唇形科（Labiatae），锥花属（Gomphostema）。产云南南部及东南部，广东和广西南部，生于海拔140-1100米的密林下潮湿处。花期4-7月。

Lustrous awl flower *(Gomphostema lucidum)* of the *Labiatae* famiy, *Gomphostema* genus, grows in wet soil in dense trees 140 to 1,100 meters above sea level in southern and southeastern Yunnan Province. Its flowering period lasts from April to July.

假糙苏（Paraphlomis japonica），唇形科（Labiatae），假糙苏属（Paraphlomis）。产云南东南部、南部及西南部，海南、台湾及广西南部，生于海拔（320-）850-1350米的热带雨林内。花期6-8月。

Paraphlomis *(Paraphlomis japonica)* of the *Labiatae* family, *Paraphlomis* genus, grows in tropical rain forests 850 to 1,350 meters above sea level in southeastern, southern and southwestern Yunnan Province, Taiwan Province and southern Guangxi Zhuang Autonomous Region. It blooms from June to August.

云南玉蕊（Barringtonia pendula），玉蕊科（Lecythidaceae），玉蕊属（Barringtonia）。产云南南部，生于海拔600-900米的密林中。花序长而下垂，雄蕊丝状，花期4-5月。

Yunnan Barringtonia *(Barringtonia pendula)* of the *Lecythidaceae* family, *Barringtonia* genus, grows in dense forests 600 to 900 meters above sea level in southern Yunnan Province. Its inflorescence is long and bends toward the ground. It blooms in April and May.

两型蜘蛛抱蛋(Aspidistra bimoda),百合科(Liliaceae),蜘蛛抱蛋属(Aspidistra)。特产广西龙虎山,生于石灰岩山常绿阔叶林下。同株植物开出的花中既有具柱头的两性花和不具柱头雄蕊的单性花。花期3-4月。

Common aspidistra (*Aspidistra bimoda*) of the *Liliaceae* family, *Aspidistra* genus, is a special local plant of Longhu Mountain in Guangxi Zhuang Autonomous Region. It grows in evergreen broadleaf forests on limestone mountains. A plant may have hermaphrodite flowers or unisexual flowers without stamen. It blooms in March and April.

巨型蜘蛛抱蛋(Aspidistra longiloba),百合科(Liliaceae)蜘蛛抱蛋属(Aspidistra)。特产广西南部十万大山地区。花大,直径达15cm,花被12-14裂,形态奇特,色彩艳丽,颇具观赏价值。花期5-6月。

Giant aspidistra *(Aspidistra longiloba)* of the *Liliaceae* family, *Aspidistra* genus, is a special local plant in the mountains in southern Guangxi Zhuang Autonomous Region. The bright-colored floral envelope of 12 to 14 petals may reach 15 centimeters in diameter in a strange form. It blooms in May and June.

罗甸蜘蛛抱蛋(Aspidistra luodianensis)百合科(Liliaceae)蜘蛛抱蛋属(Aspidistra)。特产贵州罗甸和广西天峨，生于海拔300-1000米的石灰岩山山坡或沟谷常绿阔叶林下。花大，花被8-12裂，裂片长，花未开时直立，盛开时向外反曲，形态别致，颇具观赏价值。花期6-7月。

Luodian aspidistra *(Aspidistra luodianensis)* of the *Liliaceae* family, *Aspidistra* genus, can be found only in Luodian of Guizhou Province and Tiane of Guangxi Zhuang Autonomous Region. It grows in evergreen broadleaf woods on limestone mountain slopes or valleys 300 to 1,000 meters above sea level. Its large floral envelope has 12 to 14 long petals, and stands upright before it blooms. When blooming the envelope turns outward and then curls. It is a very beautiful plant. It blooms in June and July.

嘉　兰　(Gloriosa superba)，百　合　科　(Liliaceae)，嘉兰属(Gloriosa)。产云南、广东、海南，生于海拔200-700米的平坝、低山林下或灌丛中。花期4-5月，花大单生，花形花色多姿多变，供观赏。其块茎含秋水仙碱，有剧毒。

Glorylily *(Gloriosa superba)* of the *Liliaceae* family, *Gloriosa* genus, grows in flat land, woods on hills or bushes 200 to 700 meters above sea level in the provinces of Yunnan, Guangdong and Hainan. Its single, colorful flower opens in April and May. Its stem tuber contains colchicine and is extremely poisonous.

开口箭(Tupistra Chinensis)，百合科(Liliaceae)，开口箭属(Tupistra)。产云南、广东、广西，生于海拔1000－1300米的石灰山林下。根茎药用。花期6－7月。

Chinese tupistra (*Tupistra Chinensis*) of the *Liliaceae* family, *Tupistra* genus, grows in the limestone mountain forests 1,000-1,300 meters above sea level in Yunnan, Guandong and Guangxi. Its roots and stem are of medicinal value. It blooms in July and June.

长柱开口箭(Tupistra glandistigma)，百合科(Liliaceae)，开口箭属(Tupistra)。产滇南海拔350－1600米林下。花期9－10月。

Long tupistra (*Tupistra glandistigma*) of the *Liliaceae* family, *Tupistra* genus, grows in woods below 350 to 1,600 meters above sea level. It blooms in Seprember and October.

大将军（Lobelia clavata），半边莲科（Lobeliaceae），半边莲属（Lobelia）。产云南西南部至南部，生于海拔540－2000米的山坡草地。花期2月。

Great general *(Lobelia clavata)* of the *(Lobeliaceae)* family, *(Lobelia)* genus, grows in meadows on mountain slopes 540 to 2,000 meters above sea level from southwestern to southern Yunnan Province. It blooms in February.

灰莉木(Fagraea ceilanica)，马钱科(Loganiaceae)，灰莉木属(Fagraea)。产云南、广东、广西、海南、台湾，常附生于海拔1050-1800米的石灰山季雨林中树干上。花期5月。

Fagra *(Fagraea ceilanica)* of the *Loganiaceae* family, *Fagraea* genus, grows on tree trunks in monsoon rain forests on limestone mountains 1,050 to 1,800 meters above sea level in the provinces of Yunnan, Guangdong, Hainan and Taiwan, and Guangxi Zhuang Autonomous Region, It blooms in May.

长序醉鱼草(Buddleja macrostachya)，马钱科(Loganiaceae)，醉鱼草属(Buddleja)。生于海拔1000-1300米的荒地或林缘灌丛中。花期2-3月。

Lindley butterflybush *(Buddleja macrostachya)* of the *Loganiaceae* family, *Buddleja* genus, grows in wasteland or bushes on the edge of forests 1,000 to 1,300 meters above sea level. It blooms in February and March.

绒毛含笑 (Michelia velutina)，木兰科(Magnoliaceae)，含笑属(Michelia)。产云南海拔1000米左右的山地常绿阔叶林中，西藏南部亦有分布。花期5月。用材树种，亦可供观赏。

Velvet bananashrub *(Michelia velutina)* of the *Magnoliaceae* family, *Michelia* genus, grows in evergreen broadleaf woods in the mountains about 1,000 meters above sea level in Yunnan Province and southern Tibet Autonomous Region. The timber and ornamental tree blooms in May.

云南拟单性木兰 (Parakmeria Yunnanensis)，木兰科 (Magnoliaceae)，拟单性木兰属(Parakmeria)。常绿乔木，云南特产，生于海拔1200-1800米的山谷密林中。花期5月。木材纹理颇佳。

Yunnan Parakmeria magolia *(Parakmeria Yunnanensis)* of the *Magnoliaceae* family, *Parakmeria* genus, is a evergreen tree special to Yunnan Province found in dense forests in mountain valleys 1,200 to 1,800 meters above sea level. It blooms in May. The wood has very nice veins.

金虎尾科

MALPIGHIACEAE

风筝果(Hiptage benghalensis)，金虎尾科(Malpighiaceae)，风筝果属(Hipfaga)。产云南南部至西南部，生于次生杂木林中。花期2－3月。

Kite fruit *(Hiptage benghalensis)* of the *Malpighiaceae* family, grows in secondary mixed wood from southern to southwest Yunnan Province. It blooms in February and March.

箭叶秋葵(Abelmoschus sagitifolius),锦葵科(Malvaceae),黄葵属(Abelmoschus)。产中国西南至台湾,生于海拔600-1100米的荒地,及林边潮湿处。花期9-11月。花供观赏;根药用。

Arrowleaf mallow *(Abelmoschus sagitifolius)* of the *Malvaceae* family, *Abelmoschus* genus, is found from Southwest China to Taiwan Province on wasteland or in wet land near forests 600 to 1,100 meters above sea level. Its flowering period lasts from September to November. Its flower is used for decoration and as medicine.

野牡丹(Melastoma candidum)，野牡丹科(Melastomaceae)，野牡丹属(Melastoma)。产广东、广西、台湾，生于空旷山坡上。全株入药，有解毒消肿、收敛止血之效。

Wild peony (*Melastoma candidum*) of the *Melastomaceae* family, *Melastoma* genus, grows on open mountain slopes in Guangdong Province, Taiwan Proinve and Guangxi Zhuang Autonomous Region. The whole plant can be used as medicine for treating poisoning and stopping bleeding.

尖子木(Oxyspora paniculata)，野牡丹科(Melastomaceae)，尖子木属(Oxyspora)。产云南、四川、贵州，广西，生于海拔500-1900米的山谷密林中。花期7-9月。

Oixyspor wood *(Oxyspora paniculata)* of the *Melastomaceae* family, *Oxyspora* genus, grows in dense forests in mountain valleys 500 to 1,900 meters above sea level in the provinces of Yunnan, Sichuan and Guizhou and Guangxi Zhuang Autonomous Region. Its flowering period lasts from July to September.

多花野牡丹(Melastoma polyanthum)，野牡丹科(Melastomaceae)，野牡丹属(Melastoma)。产云南、贵州至台湾各省区，生于海拔600-1300米的坡地灌丛中。灌木，全草药用。花期3-4月。

Multiflower wild peony *(Melastoma polyanthum)* of the *Melastomace*ae family, *Melastoma* genus, grows in bushes on mountain slopes in the provinces of Yunnan, Guizhou and Taiwan. Every part of the plant can be used as medicine. It blooms in March and April.

假朝天罐(Osbeckia crinita)，野牡丹科(Melastomaceae)，金锦香属(Osbeckia)。别名：罐罐花、盅盅花(以果形命名)。产云南、贵州，生于海拔1000-1300米的荒地或沼泽边缘。灌木，花期11-12月，为秋季观花植物;根叶入药，具收敛止血、活血解毒效用。

Fake skywar *(Osbeckia crinita)* of the *Melastomaceae* family, *Osbeckia* genus, also known as jar flower or cup flower, depending on the shape of the fruit of different varieties, grows on the edge of marshes or wasteland 1,000 to 1,300 meters above sea level in the provinces of Yunnan and Guizhou. The shrub blooms in November and December, a decorative plant in autumn. Its root is used to stop bleeding and release poison.

含羞草 (Mimosa pudica) 含羞草科 (Mimosaceae),含羞草属 (Mimosa)。产云南、广东、广西、海南,生于海拔200-1300米的沙滩草地。多年蔓生草本,花期7-8月。其叶被物体触及即闭合,供观赏,教学中常作试验材料。全草入药,有镇静安神之效。

Sensitive plant *(Mimosa pudica)* of the *Mimosaceae* family, *Mimosa* genus, grows on sandy meadows 200 to 1,300 meters above sea level in the provinces of Yunnan, Guangdong and Hainan and Guangxi Zhuang Autonomous Region. The perennial herb blooms in July and August. When touched the leaves will close. It is a decorative plant and an experimental plant for students. Every part of it can be used as a sedative.

眼镜豆 (Entada phaceoloides),含羞草科 (mimosaceae) 榼藤子属 (Entada)。产云南南部海拔500－1700米的低中山沟谷或密林中。花期4－5月。种子、藤皮、根皮有毒,供药用。

Eyeglass bean *(Entada phaceoloides)* of the *Mimosaceae* family, *Entada* genus, grows in mountain gullies or dense forests 500 to 1,700 meters above sea level in southern Yunnan Province. It blooms in April and May. Its seed, bark and root are poisonous but of medicinal value.

木瓜榕（Ficus auriculata），桑科（Moraceae），榕属（Ficus）。产云南南部至西南部海拔400－1000米左右的平坝、低中山杂木林中。老茎或分枝上着果，成熟可食；嫩叶可作蔬菜。花期4－5月。

Eared strangler fig *(Ficus auriculata)* of the *Moraceae* family, *Ficus* genus, grows in mixed woods on mountains 400 to 1,000 meters above sea level in southern to southwestern Yunnan Province. Fruit is born on old stems or branches. The ripe fruit and tender leaves are edible. It blooms in April and May.

歪叶榕（Ficus cyrtophylla），桑科（Moraceae），榕属（Ficus）。产云南、广西、贵州、西藏（墨脱），生于海拔(300)700-1600(2300)米的密林下或水边潮湿处。茎花植物。花期3－4月。

Tiltedleaf fig *(Ficus cyrtophylla)* of the *Moraceae* family, *Ficus* genus, grows in dense forests or near water 700 to 1,600 meters above sea level in the provinces of Yunnan and Guizhou and Guangxi Zhuang Autonomous Region, and the Medog area of Tibet Autonomous Region. Flowers grow from its stem. It blooms in March and April.

聚果榕（Ficus racemosa），桑科（Moraceae），榕属（Ficus）。产云南、贵州、广西，生于海拔500～1200米的溪边、河畔。茎花植物。5～7月开花，秋季树干上果实累累，极为壮观。

Fruitgathering fig *(Ficus racemosa)* of the *Moraceae* family, *Ficus* genus, grows by streams and rivers 500 to 1,200 meters above sea level in the provinces of Yunnan and Guizhou and Guangxi Zhuang Autonomous Region. Its flowers grow from the stem from May to July. In autumn the tree is laden with fruit, presenting a marvelous sight.

芭蕉红（Musa coccinea），芭蕉科（Musaceae），芭蕉属（Musa）。产云南东南部，生于海拔600米以下的沟谷及水分条件好的山坡上。全年均开花。花、果等均有毒，不可食用。

Japanese banana red *(Musa coccinea)* of the *Musaceae* family, *Musa* genus, grows in gullies or on mountain slopes with plenty of water below the altitude of 600 meters in southeastern Yunnan Province. Its flowering period lasts through the whole year. Both its flowers and fruit are poisonous and inedible.

粗茎紫金牛(Ardisia dasyrhigomatica)，紫金牛科(Myrsinaceae)，紫金牛属(Ardisia)。产云南南部和西部，生于海拔200-600米的密林下或沟边潮湿处。叶大，长45厘米，宽达20厘米。花期6-7月。

Thick-stem ardisia *(Ardisia dasyrhigomatica)* of the *Myrsinacea*e family, *Ardisia* genus, grows in wet land in dense forests or by ditches 200 to 600 meters above sea level in southern and western Yunnan Province. Its large leaf can reach 45 centimeters in length and 20 centimeters in width. It blooms in June and July.

水竹蒲桃(Syzygium fiuviatile),桃金娘科(Myrtaceae),蒲桃属(Syzygium)。产云南南部,生于海拔500-800米的河边沙滩及石缝中。花期10-12月,果可食。

Waterbamboo roseapple *(Syzygium fiuviatile)* of the *Myrtaceae* family, *Syzygium* genus, grows in rock crevices or on sandy river banks 500 to 800 meters above sea level in southern Yunnan Province. Its flowering period lasts from October to December. Its fruit is edible.

蒲桃(Syzygium jambos),桃金娘科(Myrtaceae),蒲桃属(Syzygium)。产台湾、广东、广西和云南,生于山溪旁灌丛中。果生食或作蜜饯。花期4-5月。

Roseapple *(Syzygium jambos)* of the *Myrtaceae* family, *Syzygium* genus, grows in bushes by mountain streams in the provinces of Taiwan, Guangdong and Yunnan and Guangxi Zhuang Autonomous Region. Its fruit can be eaten fresh or preserved with sugar.It blooms in April and May.

毛紫薇(Lagestroemia tomentosa),千屈菜科(Lythraceae),紫薇属(Lagerstroemia)。产云南南部、生于平坝及石灰岩地区海拔500-900米的疏林中。花期4-5月。

Velvet crapemystle *(Lagestroemia tomentosa)* of the *Lythraceae* family, *Lagerstroemia* genus, grows in sparse woods on plains or limestone areas 500 to 900 meters above sea level in southern Yunnan Province. It blooms in April and May.

雨林环境是附生兰的适生地，种类十分丰富。图为一株树上同时生有玫瑰石斛（Dendrobium crepidatum）、鼓槌石斛（D.chrysotoxum），报春石斛（D.primulinum）等多种兰花组成的“空中花园”。

Rain forests are an ideal habitat of many orchid varieties. On this single tree in the picture grow the rose noble dendrobium *(Dendrobium crepidatum)*, the drumstick noble dendrobium *(D. chrysotoxum)* and the springharbinger noble dendrobium *(D. primulinum)*, forming a miniature mid-air garden.

报春石斛（Dendrobium primulinum），产云南南部，生于海拔700-1800米的疏林中树干上。花期3-4月。

Spring Dendrobium (*Dendrobium primulinum*) grows on tree trunks in the sparse forests 700-1,800 meters above sea level in south Yunnan Province. It blooms in March and April.

鼓槌石斛（Dendrobium chrysotoxum），兰科（Orchidaceae），石斛属（Dendrobium）。产云南西南部及南部，生于海拔500-1500米的常绿阔叶林内，附生于树上。茎丛生，呈棒状或纺槌形，故此得名。花期4－5月。

Drumstick noble dendrobium *(Dendrobium chrysotoxum)* of the *Orchidaceae* family, *Dendrobium* genus, grows on evergreen broad leaf trees 500 to 1,500 meters above sea level in southwestern and southern Yunnan Province. The stems in clusters resemble a stick or a spindle, thus its name. The plant blooms in April and May.

兜唇石斛（Dendrobium aphyllum），兰科（Orchidaceae），石斛属（Dendrobium）。分布于广西、云南、贵州海拔400-1,500米的林中树上。花期3-4月。

Apron-lip dendrobium(*Dendrobium aphyllum*) grows on trees in the forests 400-1,500 meters above sea level in Guanxi, Yunnan and Guizhou. It blooms in March and April.

宿苞兰(Cryptochilus luteus)，兰科(Orchidaceae)，宿苞兰属(Cryptochilus)。产云南东南部、南部至西南部及西藏东南部(墨脱)，附生在海拔1000－2300米密林中树上。5月开花，花密集，总状花序排为两列，甚雅致。

Cryptoch orchid *(Cryptochilus luteus)* of the *Orchidaceae* family, *Cryptonchilus* genus, grows on trees in dense forests 1,000 to 2,300 meters above sea level in southeastern, southern and southwestern Yunnan Province and in the Medog area in southeastern Tibet Autonomous Region. It blooms in May. Elegant blossoms open closely on the two flanks of the main inflorescence.

束花石斛(Dendrobium chysanthum)，兰科(Orchidaceae)，石斛属(Dendrobium)。产云南南部，附生于海拔800-1200米的林中树上或岩石上。花期10-11月。

Clustered-flower herba dendrobii *(Dendrobium chysanthum)* of the *Orchidaceae* family, *Dendrobium* genus, grows on trees or rocks in woods 800 to 1,200 meters above sea level in southern Yunnan Province. It blooms in October and November.

曲轴石斛(Dendrobium gibsonii),兰科(Orchidaceae),石斛属(Dendrobium)。产云南南部、广西,生于海拔700-900米的林缘树上或岩石上。花期6-7月。茎药用。

Curved-axis herba dendrobii *(Dendrobium gibsonii)* of the *Orchidaceae* family, *Dendrobium* genus, grows on trees or rocks in woods 700 to 900 meters above sea level in southern Yunnan Province and Guangxi Zhuang Autonomous Region. It blooms in June and July.

肿节石斛(Dendrobium pendulum),兰科(Orchidaceae),石斛属(Dendrobium)。产云南西南部至南部,附生于海拔1000-1800米的林中树上。花期3-5月。

Swollen-joint herba dendrobii *(Dendrobium pendulum)* of the *Orchidaceae* family, *Dendrobium* genus, grows on trees in woods 1,000 to 1,800 meters above sea level in southwestern and southern Yunnan Province. It blooms from March to May.

球 花 石 斛 (Dendrobium thysiflorium), 兰 科 (Orchidaceae), 石 斛 属 (Dendrobium)。产中国西南大部分省区，广东亦产，生于海拔 760-2400 米的热带林中的树上或石上。总状花序由茎顶抽出，密集下垂。花期 4-6 月。

Ball-flower herba dendrobii *(Dendrobium thysiflorium)* of the *Orchidaceae* family, *Dendrobium* genus, grows on trees or rocks in tropical forests 760 to 2,400 meters above sea level in Guangdong Province and most part of Southwest China. The main inflorescence sprouts from the top of the stem and bends downward. It blooms from April to June.

爪哇毛兰(Eria javanica)，兰科(Orchidaceae)，毛兰属(Eria)。产云南南部，生于海拔670-1000米的林下岩石上。花白色，极芳香，花期6-7月。

Java velvet orchid *(Eria javanica)* of the *Orchidaceae* family, *Eria* genus, grows on rocks under trees 670 to 1,000 meters above sea level in southern Yunnan Province. Its white, fragrant flowers open in June and July.

紫花美冠兰（Eulophia nuda），兰科（Orchidaceae），美冠兰属（Eulophia）。生于海拔850米的疏林下。花美丽，花期5-6月。

Purple flower orchid (*Eulophia nuda*) of the *Orchidaceae* family, *Eulophia* genus, is found in sparse woods 850 meters above sea level. Its beautiful flowers open in May and June.

多花指甲兰(Aerides rosea)，兰科(Orchidaceae)，指甲兰属(Aerides)。产云南南部和广西，附生于海拔900-1300米的热带季雨林树上。花为玫瑰色，密集、壮观，令人赏心悦目，花期6 - 7月。

Fingernail orchid *(Aerides rosea)* of the *Orchidaceae* family, *Aerides* genus, grows on trees in tropical monsoon rain forests 900 to 1,300 meters above sea level in southern Yunnan Province and Guangxi Zhuang Autonomous Region. In June and July its rose-color blossoms open in close clusters, very pleasant to the eye.

滇 坛 花 兰（Acanthephippium sylhetense），兰科（Orchidaceae），坛花兰属（Acanthephippium）。产云南南部，生于海拔700-900米的热带雨林中。地生兰，花期3-4月。

Yunnan orchid *(Acanthephippium sylhetense)* of the *Orchidaceae* family, *Acanthephippium* genus, grows on the ground in tropical rain forests 700 to 900 meters above sea level in southern Yunnan Province. It blooms in March and April.

蜘蛛兰（Arachnis labrosa），兰科（Orchidaceae），蜘蛛兰属（Arachnis）。产广东、广西、云南，附生于海拔800-1200米的树上。因其花形如蜘蛛而得名，花期2-3月。

Narrowlip spider orchid *(Arachnis labrosa)* of the *Orchidaceae* family, *Arachnis* genus, grows on trees 800 to 1,200 meters above sea level in the provinces of Guangdong and Yunnan and Guangxi Zhuang Autonomous Region. Its flowers which open in February and March resemble spiders, thus its name.

乌舌兰(Ascocentrum ampullaceum),兰科(Orchidaceae),乌舌兰属(Ascocentrum)。产云南西双版纳,生于海拔900-1100米的常绿阔叶林中,附生于树干上。花期4-5月。

Bird's tongue orchid *(Ascocentrum ampullaceum)* of the *Orchidaceae* family, *Ascocentrum* genus, grows on trees in evergreen, broadleaf woods 900-1,100 meters above sea level in Xishuangbanna, Yunnan Province. It blooms in April and May.

直唇石豆兰(Bulbophyllum delitescens),兰科(Orchidaceae),石豆兰属(Bulbophyllum)。产云南、西藏墨脱、广东沿海地区,附生于海拔750-900米溪旁岩石或树上。花期3-4月。

Straight-lip stonebean orchid *(Bulbophyllum delitescens)* of the *Orchidaceae* family, *Bulbophyllum* genus, grows on rocks or trees by streams 750 to 900 meters above sea level in Yunnan Province, along the coast of Guangdong Province and in the Medog area of Tibet Autonomous Region. It blooms in March and April.

三褶虾脊兰(Calanthe triplicata)，兰科(Orchidaceae)，虾脊兰属(Calanthe)。产云南南部及西南部，生于海拔600 - 900米的林下沟边潮湿的环境中。花期4 - 6月。

Three-fold calanthe orchid *(Calanthe triplicata)* of the *Orchidaceae* family, *Calanthe* genus, grows in wet spots in forests 600-900 meters above sea level in southern and southwestern Yunnan Province. Its flowering period lasts from April to June.

异唇兰(Chiloschista yunnanensis)，兰科(Orchidaceae)，异唇兰属(Chiloshita)。产云南东南部、南部和西南部，附生于海拔1100-1800米的疏林中树上，特别是西双版纳的老茶树上为最多。花期6-7月。

Abnormal-lip orchid *(Chiloschista yunnanensis)* of the *Orchidaceae* family, *Chiloshita* genus, grows on trees in sparse woods 1,100 to 1,800 meters above sea level in southeastern, southern and southwestern Yunnan Province. A great number of it is found on old tea trees in Xishuangbannan of Yunnan Province. It blooms in June and July.

冬凤兰(Cymbidium dayanum)，兰科(Orchidaceae)，兰属(Cymbidium)。产华南、西南，生于海拔300-1,600米的林中树上。或枯朽树上。花期11月至翌年1月。

Winter phoenix orchid *(Cymbidium dayanum)* of the *Orchidaceae* family, *Cymbidium* genus, grows on trees or withered trees in the forests 300-1,600 meters above sea level in south and southwest China. The flowering period lasts from November to January of the next year.

硬叶吊兰(Cymbidium bicolor)，兰科(Orchidaceae)，兰属(Cymbidium)。产云南、广东、广西，附生于海拔700-1200米的树上和岩石上。花序下垂，花期4-5月。

Hardleaf hanging orchid *(Cymbidium bicolor)* of the *Orchidaceae* family, *Cymbidium* genus, grows on trees or rocks 700 to 1,200 meters above sea level in the provinces of Yunnan and Guangdong and Guangxi Zhuang Autonomous Region. The inflorescence bends downward. It blooms in April and May.

假万带兰(Vandopsis gigantea)，兰科(Orchidaceae)，假万带兰属(Vandopsis)。产广西、云南南部，生于海拔800-1,700米的山地林缘或疏林中大树上，花期3－4月。

Gigantic falsevanda *(Vandopsis gigantea)* of the *Orchidaceae* family, *Vandopsis* genus, grows on the edge of the mountain forests or on large trees in sparse forests 800-1,700 meters above sea level in Guangxi and south Yunnan. It blooms in March and April.

文山红柱兰(Cymbidium wenshanense)，兰科(Orchidaceae)，兰属(Cymbidium)。特产云南东南部，附生于海拔1100-1300米的常绿阔叶林树上或林下石上。合蕊柱顶端为血红色，其余白色，美丽，花期1-3月。

Wenshan redpillar orchid *(Cymbidium wenshanense)* of the *Orchidaceae* family, *Cymbidium* genus, grows under trees or rocks in evergreen broadleaf forests 1,100 to 1,300 meters above sea level in southeastern Yunnan Province. It blooms from January to March.

阔叶沼兰（Malaxia iatifolia），兰科（Orchidaceae），沼兰属（Malaxia）。产云南、广东、广西、台湾，生于海拔900-1900米的林缘或林下阴湿处。花期5-6月。

Broadleaf marsh orcid *(Malaxia iatifolia)* of the *Orchidaceae* family, *Malaxia* genus, grows in wet land in woods or on the edge of forests 900 to 1,900 meters above sea level in the provinces of Yunnan, Guangdong and Taiwan and Guangxi Zhuang Autonomous Region. It blooms in May and June.

凤蝶兰（Papilionantha teres），兰科（Orchidaceae），凤蝶兰属（Papilionantha）。产云南，生于海拔600-800米的疏林中，附生于树上。花大美丽，是一种极佳的热带花卉，花期3-4月。

Butterfly orchid *(Papilionantha teres)* of the *Orchidaceae* family, *Papilionantha* genus, grows on trees in sparse woods 600 to 800 meters above sea level in Yunnan Province. Its big and beautiful flowers open in March and April.

巨瓣兜兰(Paphiopedilum bellatulum),兰科(Orchidaceae),兜兰属(Paphiopedilum)。产云南东南部,生于海拔1000-1800米石灰岩缝隙中荫蔽、腐殖质丰富、苔藓多的土壤上。2-5月开花。花瓣巨大,中萼片、花瓣内侧和唇瓣上布满紫红色或褐色斑点,美丽。

Giantpetal orchid (*Paphiopedilum bellatulum*) of the *Orchidaceae* family, *Paphiopedilum* genus, grows in limestone crevices with rich humu soil and moss 1,000 to 1,800 meters above sea level in southern Yunnan Province. It blooms from February to May. On the sepals and the inside of large petals are beautiful purple or brown specks.

同色兜兰(Paphiopedilum concolor),又名黄花兜兰,兰科(Orchidaceae),兜兰属(Paphiopedilum)。产云南和广西,生于海拔1430米以下的低山沟谷林中。花期4-5月。

Unicolor Orchid (*Paphiopedilum concolor*) of the *Orchidacea*e family, *Paphiopedilum* genus, grows in woods in mountain valleys lower than 1,430 meters above sea level in Yunnan Province and Guangxi Zhuang Autonomous Region. It blooms in April and May.

长瓣兜兰(Paphiopedilum dianthum),兰科(Orchidaceae),兜兰属(Paphiopedilum)。中国特有种,产贵州、广西和云南南部,生于海拔1000-1800米的常绿阔叶林树干上或岩石覆土上。7-9月开花。花具1对长而扭曲的花瓣,好似仙女下凡甩向空中的飘带。

Longpetal orchid *(Paphiopedilum dianthum)* of the *Orchidaceae* family, *Paphiopedilum* genus, grows on evergreen broadleaf trees or in the soil on rocks 1,000 to 1,800 meters above sea level in Guizhou Province, Guangxi Zhuang Autonomous Region and southern Yunnan Province. The special orchid variety found only in China blooms from July to September. Each blossom has a pair of twisted long petals like silk ribbons falling down from the sky.

紫毛兜兰(Paphiopedilum Villosum),兰科(Orchidaceae),兜兰属(Paphiopedilum)。产云南和广西,生于海拔1000米以下林中阴湿的腐殖土上。花期11－3月。本种因观赏价值高,已于18世纪就被引种至欧洲。

Purple-down orchid *(Paphiopedilum villosum)* of the *Orchidaceae* family, *Paphiopedilum* genus, grows in wet humu soil in woods lower than 1,000 meters above sea level in Yunnan Province and Guangxi Zhuang Autonomous Region. It blooms from November to March of the next year. This highly decorative variety was introduced to Europe in the 18th century.

晶帽石斛 (Dendrobium crystallinum)，兰科 (Orchidaeae)，石斛属 (Dendnobium)。产云南西双版纳，生于海拔900 - 1100米的常绿阔叶林中，附生于树上。花期6 - 7月。

Crystal-capped noble dendrobium (*Dendrobium cyrstallinum*) of the *Orchidaceae* family, *Dendrobium* genus, grows in evergreen broadleaf woods 900-1,100 meters above sea level in Xishuangbanna, Yunnan Province. It blooms in June and July.

麻栗坡兜兰 (Paphiopedilum malipoens)，兰科 (Orchidaceae)，兜兰属 (Paphiopedilum)。产云南东南部海拔1200-1500米的石灰岩混交林下沃土上。花期4 - 5月

Masubo orchid *(Paphiopedilum malipoens)* of the *Orchidaceae* family, *Paphiopedilum* genus, grows in fertile soil in mixed woods on limestone mountains 1,200 to 1,500 meters above sea level in southeastern Yunnan Province. It blooms in April and May.

硬叶兜兰(Paphiopedilum micranthum)，兰科(Orchidaceae)，兜兰属(Paphiopedilum)。中国特有种，产云南东南部和贵州西南部，地生或半地生于海拔1000-1500米的疏林中，沟边或岩石上。花冠直径7-8厘米，十分美丽，花期4-5月。

Hardleaf orchid *(Paphiopedilum micranthum)* of the *Orchidaceae* family, *Paphiopedilum* genus, is a special orchid variety of China. It grows by ditches or on rocks in sparse woods 1,000 to 1,500 meters above sea level in southeastern Yunnan Province and southwestern Guizhou Zhuang Autonomous Region. Its extremely beautiful corolla is 7-8 centimeters in diameter. It blooms in April and May.

彩云兜兰 (Paphiopedilum wardii)，兰科 (Orchidaceae)，兜兰属 (Paphiopedilum)。产云南西南部，生于海拔1200-1700米多石的土壤上或苔藓和腐殖质丰富的岩石上。花期12月至次年3月。

Colorcloud orchid *(Paphiopedilum wardii)* of the *Orchidaceae* family, *Paphiopedilum* genus, grows in rocky soil or on rocks with plenty of moss and humu soil 1,200 to 1,700 meters above sea level in southwestern Yunnan Province. It blooms from December to March next year.

长茎鹤顶兰 (Phaius longicruris)，兰科 (Orchidaceae)，鹤顶兰属 (Phaius)。产云南西双版纳海拔1400米左右林下。花期6－8月。

Longstem cranecrown orchid *(Phaius longicruris)* of the *Orchidaceae* family, *Phaius* genus, grows under trees 1,400 meters above sea level in Xishuangbanna, Yunnan Province. Its flowering period lasts from June to August.

泰香果兰（Vanilla siamensis），兰科（Orchidaceae），香果兰属（Vanilla）。产云南南部，附生于海拔800-1000米的热带密林中树上或石头上，为香果兰的野生类型。花黄绿色，叶肉质，有较高的观赏价值。花期11-12月。

Thai Vanilla *(Vanilla siamensis)* of the *Orchidaceae* family, *Vanilla* genus, grows on trees or rocks in dense tropical forests 800 to 1,000 meters above sea level in southern Yunnan Province. Its yellowish green flowers open in November and December. With its fleshy leaves

虎斑兜兰（Paphiopedilum markianum）。兰科（Orchidaceae），兜兰属（Paphiopedilum）。近年发现的新种。产云南西部，生于海拔1200-2200米的灌木或疏林山坡上湿润或稍荫蔽的地方。6-8月开花。花大，直径达8-9cm，在绿黄色的中萼片和前部为淡紫色的花瓣上有3条紫褐色的斑纹，艳丽别致，有较高的观赏价值。花期9-10月。

Tigerspot orchid *(Paphiopedilum markianum)* of the *Orchidaceae* family, Paphiopedilum genus, is a new variety discovered only in recent years in western Yunnan Province. It grows in wet land under bushes or in sparse woods 1,200 to 2,200 meters above sea level. Its flowers, 8-9 centimeters in diameter, open from June to August. Three purple stripes appear on the greenish yellow sepals and petals with light purple tips. The plant is of high decorative value.It blooms in September and October.

紫纹兜兰(Paphiopedilum purpuratum)，兰科(Orchidaceae)，兜兰属(Paphiopedilum)。又名香港兜兰。产广东南部、香港和云南东南部，生于海拔 50-700 米杂草中或疏林山坡上多石的地方。此种植物因其花色艳丽华贵和原产香港之故，被人们称为“香港小姐”。花期 10 月至次年 1 月。花直径 7-8cm。

Purplevein orchid *(Paphiopedilum purpuratum),* also named Hongkong orchid, grows among weeds or in sparse woods on rocky mountain slopes 50 to 700 meters above sea level in southern Guangdong Province, Hong Kong and southeastern Yunnan Province. It is a native plant of Hong Kong and very beautiful. It is commonly called "Miss Hong Kong". Its flowering period lasts from October to January the following year.

竹叶兰(Arundina graminifolia)，兰科(Orchidaceae)，竹叶兰属(Arundina)。产云南、广西、广东、海南等省、区海拔800-1,300米的山坡草地和疏林中。花期3-10月。

Chinese arundina *(Arundina graminifolia)*of the *Orchidaceae* family, *Arundina* genus, grows on the slope grassland and in the sparse forests 800-1,300 meters above sea level in Guangxi, Guangdong and Hainan. Its flowering period lasts from March to October.

杓唇石斛(Denrobium moschatum)，兰科(Orchidaceae)，石斛属(Denrobium)。产云南南部，生于海拔1,300米的疏林中树上。花期7-8月。

Spoon-lip denrobium *(Denrobium moschatum)* of the *Orchidaceae* family, *Denrobium* genus, grows on trees in the dense forests 1,300 meters above sea level in Yunnan Province. It blooms in July and August.

白点兰(Thrixspermum centipeda)，兰科(Orchidaceae)，白点兰属(Thrixspermum)。产云南南部、广东、海南等省，附生于海拔700-900米的疏林中树上。花期2-3月。

Whitespeck orchid *(Thrixspermum centipeda)* of the *Orchidaceae* family, *Thrixspermum* genus, grows on trees in sparse woods 700 to 900 meters above sea level in southern Yunnan Province, and Guangdong and Hainan provinces. It blooms in February and March.

笋兰(Thunia alba)，兰科(Orchidaceae)，笋兰属(Thunia)。产云南、四川、西藏等省、区，生于海拔900-2000米的石山阔叶林下，附生于岩石或树上。因其茎如笋而得名。花体硕大，美丽而芳香，5-6月开花。全草入药，能清肺止咳。

Bambooshoot orchid (*Thunia alba*) of the *Orchidaceae* family, *Thunia* genus, grows on trees or rocks in broadleaf woods on rocky mountains 900 to 2,000 meters above sea level in the provinces of Yunnan and Sichuan and Tibet Autonomous Region. It blooms in May and June. Every part of the plant is of medicinal value, good for treating coughing.

大花万带兰(Vanda coerulea)，兰科(Orchidaceae)，万带兰属(Vanda)。产云南西双版纳，生于海拔1150-1600米的林中或林缘树上。花大而美丽，花期11-12月。

Bigflower vanda orchid *(Vanda coerulea)* of the *Orchidaceae* family, *Vanda* genus, grows under or on trees 1,150 to 1,600 meters above sea level in Xishuangbanna, Yunnan Province. Its large, beautiful flowers open in November and December.

小花万带兰(Vanda coerulescens)，兰科(Orchidaceae)，万带兰属(Vanda)。产台湾、广东、广西、海南及云南南部，附生于海拔500 - 800米的低山沟谷密林中树上。花期4 - 5月。

Smallflower vanda orchid *(Vanda coerulescens)* of the *Orchidaceae* family, *Vanda* genus, grows on trees in dense forests in mountain gullies 500 to 800 meters above sea level. It blooms in April and May.

大花鹤顶兰(Phaius magniflorus)，兰科(Orchidaceae)，鹤顶兰属(Phaius)。产云南西双版纳，生于海拔900-1100米的密林阴湿处。花期4-5月。

Largeflower cranecrown orchid *(Phaius magniflorus)* of the *Orchidaceae* family, *Phaius* genus, grows in wet land in dense forests 900 to 1,100 meters above sea level in Xishuangbanna, Yunnan Province. It blooms in April and May.

黄花鹤顶兰（Phaius flavus），兰科（Orchidaceae），鹤顶兰属（Phaius）。产云南南部，生于海拔900米以下的林下阴湿处。花期8-9月，供观赏。

Yellow flower cranecrown orchid *(Phaius flavus)* of the *Orchidaceae* family, *Phaius genus*, grows in wet land under trees at an elevation of lower than 900 meters in southern Yunnan Province. Its decorative flowers open in August and September.

五膜草 (Pentaphragma sinensis)，五膜草科 (Pentaphragmaceae)，五膜草属 (Pentaphragma)。产云南西双版纳，生于海拔500-700米的林下阴湿处。花期7月。

Pentaphragma *(Pentaphragma sinensis)* of the *Pentaphragmaceae* family, *Pentaphragma* genus, grows in shady and wet spots in woods 500-700 meters above sea level in Xishuangbanna, Yunnan Province. It blooms in July.

单穗鱼尾葵(Caryota monostachys)，棕榈科(Palmae)，鱼尾葵属(Caryota)。产云南南部，生于海拔700-900米的热带密林中。花序修长、别致，作观赏、绿化用。花期4-5月。

Fishtailpalm *(Caryota monostachys)* of the *Palmae* family, *Caryota* genus, grows in dense tropical forests 700 to 900 meters above sea level in southern Yunnan Province. Its inflorescence is very long and its beautiful flowers open in April and May.

青棕(Caryota ochlandra)，棕榈科(Palmae)，鱼尾葵属(Caryota)。产云南、广东、广西、海南、福建等省区，生于海拔450-900米的山坡或沟谷林中。花期5-7月。

Green palm *(Caryota ochlandra)* of the *Palmae* family, *Caryota* genus, grows in woods on mountain slopes or in gullies 450 to 900 meters above sea level in the provinces of Yunnan, Guangdong, Hainan and Fujian, and Guangxi Zhuang Autonomous Region. It blooms from May to July.

蓟罂粟(Argemone mexicana),罂粟科(Papaveraceae),刺罂粟属(Argemone)。产云南西双版纳、保山怒江坝、东川金沙江海拔850-1200米的田坝或江边,广东沿海有逸为野生。花供观赏,种子含油30%左右,黄色汁液含生物碱(Argemonine)。花期6-7月

Mexican pricklepoppy *(Argemone mexicana)* of the *Papaveraceae* family, *Argemone* genus, grows in fields or by the rivers 850 to 1,200 meters above sea level in Xishuangbanna, by the Nu River in Baoshan and by the Jinsha River in Dongchuan of Yunnan Province. Some can also be found along the coast in Guangdong Province. The decorative plant blooms in June and July. Its seeds contain 30 percent of oil. Argemonine can be extracted from its yellow Juice. It blooms in June and July.

舞草(Codariocalyx motorium),又名"风流草",蝶形花科(Papilionaceae),山蚂蝗属(Codariocalyx)。产云南(西双版纳)、广西、台湾,生于海拔900-1200米的山坡次生林及杂木林中。小灌木,当人们唱歌时,受声波感应,小叶会随之摆动,故而得名。花期10-11月。

Dancing grass *(Codariocalyx motorium),* also known as "elegant grass", is of the *Papilionaceae* family, *Codariocalyx* genus, grows in secondary or mixed woods on mountain slopes 900 to 1,200 meters above sea level in Xishuangbanna of Yunnan Province, Guangxi Zhuang Autonomous Region and Taiwan Province. When people sing its tiny leaves swing, thus its name. It blooms in October and November.

黄雀儿(Crotalaria cytisoides),蝶形花科(Papilionaceae),野百合属(Crotalaria)。产云南南部至东南部,生于海拔600-1200米的林缘河边及灌丛中。花供观赏,花期9-10月。

Siskin wild lily *(Crotalaria cytisoides)* of the *Papilionaceae* family, *Crotalaria* genus, grows near rivers or in bushes on the edge of forests 600 to 1,200 meters above sea level in southern and southeastern Yunnan Province. Its ornamental flowers open in September and October.

劲直刺桐,(Erythrina stricata),蝶形花科(Papilionaceae),刺桐属(Erythrina)。产云南,生于海拔800-1500米的疏林中,村寨附近时有栽培。花期4-5月。

Straight coralbean *(Erythrina stricata)* of the *Papilionaceae* family, *Erythrina* genus, grows in sparse woods near villages 800 to 1,500 meters above sea level in Yunnan Province. Sometimes people plant it near villages. It blooms in April and May.

大果油麻藤(Mucuna macrocarpa),蝶形花科(Papilionaceae),藜豆属(Mucuma)。产云南、海南,生于海拔500-1500米的林缘,多年生藤本,常攀附于灌木上。花期4-6月。

Prickly velvetbean *(Mucuna macrocarpa)* of the *Papilionaceae* family, *Mucuma* genus, grows at the edge of forests 500 to 1,500 meters above sea level in Yunnan and Hainan provinces. The perennial vine often climbs up bushes and blooms from April to June.

海 南 油 麻 藤 (Mucuma　hainanensis), 蝶 形 花 科 (Papilionaceae), 藜豆属(Mucuma)。产云南南部和海南, 生于林缘、河边灌丛中。花期 11-12 月。

Hainan mucuna *(Mucuna hainanensis)* of the *Papilionaceae* family, *Mucuna* genus, grows on the edge of forests or in bushes by rivers in southern Yunnan Province and Hainan Province. It blooms in November and December.

间序油麻藤(Mucuma interrupta),蝶形花科(Papilionaceae) 藜豆属(Mucuma)。产云南、海南,生于海拔600-900米的林中。花期11月。

Interrupted-inflorescence mucuna *(Mucuma interrupta)* of the *Papilionaceae* family, *Mucuma* genus, grows in woods 600 to 900 meters above sea level in the provinces of Yunnan and Hainan. It blooms in November.

马 鹿 花 (Butea subercta)，蝶 形 花 科 (Papilionace)，紫 铆 属 (Butea)。生于云南南部 900-1100 米的林中。花期 11-12 月。

Horse-Deer Butea *(Butea subercta)* of the *Papilionaceae* family, *Pueraria* genus, grows in woods 900 to 1,100 meters above sea level in southern Yunnan Province. It blooms in November and December.

猫尾射(Uraria crinita)，蝶形花科(Papilionaceae)，狸尾豆属(Uraria)。产云南、广东、广西、海南，生于海拔 600-700 米的林缘荒地或灌丛中。花序直立，美丽，供观赏，因花形如猫尾而得名。根药用，花期 4-5 月。

Cattail Uraria *(Uraria crinita)* of the *Papilionaceae* family, *Uraria* genus, grows in wasteland or bushes on the edge of forests 600 to 700 meters above sea level in the provinces of Yunnan, Guangdong and Hainan and Guangxi Zhuang Autonomous Region. Its inflorescence stands upward and its flowers which open in April and May look like the cat's tail, thus its name.

长柱山丹(Duperrea pavettaefolia),茜草科(Rubiaceae),长柱山丹属(Duperrea)。产云南南部,生于海拔500-900米的密林下或林缘灌丛中。花期6月。

Long-tube morningstar lily *(Duperrea pavettaefolia)* of the *Rubiaceae* family, *Duperrea* genus, grows in dense forests or in bushes on the edge of forests 500-900 meters above sea level in southern Yunnan Province. It blooms in June.

小仙丹花(Ixora henryi),茜草科(Rubiaceae),龙船花属(Ixora)。产广西、云南南部海拔500-900米的沟谷密林下。全株药用。花期5-6月。

Small Ixora *(Ixora henryi)* of the *Rubiaceae* family, *Ixora* genus, grows in gullies in dense forests 500-900 meters above sea level in Guangxi Zhuang Autonomous Region and southern Yunnan Province. The whole plant is of medicinal value and it blooms in May and June.

滇仙丹花(Ixora yunnanensis),茜草科(Rubiaceae),龙船花属(Ixora)。产云南南部,生于海拔900-1100米的林下。花期6-7月,花状密集,可作庭园观赏。

Yunnan Ixora *(Ixora yunnanensis)* of the *Rubiaceae* family, *Ixora* genus, grows under trees 900 to 1,100 meters above sea level in southern Yunnan Province. Its inflorescences are very close. It blooms in June and July. The beautiful plant is often used to decorate courtyards.

馥郁滇丁香(Luculia gratissima),茜草科(Rubiaceae),滇丁香属(Luculia)。产西藏东南部和云南,生于海拔800-2000米的常绿阔叶林下。花期8-9月。

Fragrant luculia *(Luculia gratissima)* of the *Rubiaceae* family, *Luculia* genus, grows in evergreen broadleaf woods 800 to 2,000 meters above sea level in southeastern Tibet Autonomous Region and Yunnan Province. It blooms in August and September.

狭叶巴戟(Morinda angustifolia),茜草科(Rubiaceae),巴戟天属(Morinda)。产云南,生于海拔600-900米的中低山次生林下。根可提取黄色染料,亦作药用,可治黄疸型肝炎。花期4-5月,供观赏。

Indianmulberry *(Morinda angustifolia)* of the *Rubiaceae* family, *Morinda* genus, grows in secondary woods on mountains 600 to 900 meters above sea level in Yunnan Province. The yellow juice extracted from its root can be used as a dye and its root is used as medicine for treating ictero-type hepatitis. The ornamental plant blooms in April and May.

红毛玉叶金花(Mussaenda hossei),茜草科(Rubiaceae),玉叶金花属(Mussaenda)。产云南,生于海拔700-900米的次生林灌丛中,4-5月开金色小花,有白色叶状苞片吸引昆虫传播花粉,故得玉叶金花之名。

Redvelvet jadeleaf goldenflower *(Mussaenda hossei)* of the *Rubiaceae* family, *Mussaenda* genus, grows in bushes among secondary woods 700 to 900 meters above sea level in Yunnan Province. Its tiny white blossoms open in April and May. The white leaf-like budding petals attract insects, thus its name: jadeleaf goldenflower.

弯管花(Chassalia curviflora)，茜草科(Rubiacaea)，弯管花属(Chassalia)。产中国西南部，生于海拔700-900米的林下。花期6-7月。根药用。

Curved-tube flower *(Chassalia curviflora)* of the *Rubiaceae* family, *Chassalia* genus, grows under trees 700 to 900 meters above sea level in Southwest China. It blooms in June and July and its root is of medicinal value.

广东玉叶金花(Mussaenda kwangtungensis)，茜草科(Rubiaceae)，玉叶金花属(Mussaenda)。特产广东南部，生于海拔200-400米的山地丛林中，常攀援于林冠上。花期6-7月

Guangdong jadeleaf goldenflower *(Mussaenda kwangtungensis)* of the *Rubiaceae* family, *Mussaenda* genus, can be found only in woods on mountains 200-400 meters above sea level in southern Guangdong Province. It often climbs up the crown of a tree. It blooms in June and July.

裂果金花（Schizomussaenda dehiscens），茜草科（Rubiaceae），裂果金花属（Schizomussaenda）。产云南和广西南部，生于海拔700-900米的山坡次生林中。花期3-4月。根药用，具消热解毒、消炎利尿之功效。

Splitfruit goldenflower *(Schizomussaenda dehiscens)* of the *Rubiaceae* family, *Schizomussaenda* genus, grows in secondary woods on mountains 700 to 900 meters above sea level. It blooms in March and April and its root is used as medicine for treating inner heat, inflammation and urinating difficulty.

大叶钩藤（Unacaria macrophylla），茜草科（Rubiaceae），钩藤属（Unacaria）。产广东、广西、云南，生于海拔600-900米的林中，攀援于树上。花期8-9月。钩供药用，有镇静功效。

Gambplant *(Unacaria macrophylla)* of the *Rubiaceae* family, *Unacaria* genus, grows in forests 600 to 900 meters above sea level in the provinces of Guangdong and Yunnan and Guangxi Zhuang Autonomous Region. The vine climbs up tree crowns to live and blooms in August and September. Its hook is used as a sedative.

九里香(Murraya paniculata),芸香科(Rutaceae),九里香属(Murraya)。产台湾、广东、广西、云南南部,生于海拔700-800米的林中或石灰岩山地。其根和叶药用,有行气镇痛,消肿活络功效,可治感冒头痛、牙痛、胃痛、风湿骨痛等症。花期4-8月。

Nine-mile fragrance *(Murraya paniculata)* of the *Rutaceae* family, *Murraya* genus, grows in woods or on limbstone mountains 700 to 800 meters above sea level in the provinces of Taiwan and Guangdong, Guangxi Zhuang Autonomous Region, and southern Yunnan Province. Its root and leaf can be used as medicine to relieve headache due to cold, teethache, pain in stomach and pain caused by arthritis, and abate the swelling. It blooms from April to August.

紫萼蝴蝶草（Torenia vioacea），玄参科（Scrophulariaceae），蝴蝶草属（Torenia）。产中国南部和西南部，生于海拔800-1200米的山坡路边草地或林缘草丛中。花期7-8月，供观赏。

Purplesepal butterfly *(Torenia vioacea)* of the *Scrophulariaceae* family, *Torenia* genus, grows on mountain slopes, in grass by the road or on the edge of woods 800 to 1,200 meters above sea level in southern and Southwest China. Its beautiful flowers open in July and August.

八宝树（Duabanga grandiflora），海桑科（Sonneratiaceae），八宝树属（Duabanga）。产云南，生于海拔500-1100米的密林边缘或河边潮湿处。花期4-5月。乔木用材。

Eight-treasure tree *(Duabanga grandiflora)* of the *Sonneratiaceae* family, *Duabanga* genus, grows on the edge of dense forests or wet land by the river 500 to 1,100 meters above sea level in Yunnan Province. The timber tree blooms in April and May.

翅子树(Pterospermum acerifolium),梧桐科(Sterculiaceae),翅子树属(Pterospermum)。产云南南部,生于海拔800-1300米的常绿阔叶林中或沟谷林中;用材,树皮药用;纤维供搓绳。花期7月。

Wingleaf tree *(Pterospermum acerifolium)* of the *Sterculiaceae* family, *Pterospermum* genus, grows in evergreen broadleaf woods or woods in gullies 800-1,300 meters above sea level in southern Yunnan Province. Bark of this timber tree can be used as medicine and its fiber can be woven into ropes which are preserved as seeds of the tree. It biooms in July.

火筒(Firmiana colorata),梧桐科(Sterculiaceae),梧桐属(Firmiana)。产云南南部及东南部,生于海拔120-950(-1100)米的热带密林中或石灰岩石缝中。花期3-4月,开花时满树皆红,可作庭园绿化观赏。

Firetube *(Firmiana colorata)* of the *Sterculiaceae* family, *Firmiana* genus, grows in dense tropical rain forests or in limestone crevices 120 to 950 meters above sea level in southern and southeastern Yunnan Province. The tree crown is covered with red flowers in March and April, a very beautiful tree to decorate the courtyard.

老虎须（Tacca chantrieri），箭根薯科（Taccaceae），箭根薯属（Tacca）。产云南、广东、广西，生于海拔500-900米雨林下的水边阴湿处。因其条形苞片似虎须而得名。花期4-5月。根茎药用，治胃肠溃疡、高血压、肝炎，外敷治烧伤、烫伤等。有小毒，慎用。

Tigerwhisker *(Tacca chantrieri)* of the *Taccaceae* family, *Tacca* genus, grows in wet land in rain forests or by water 500 to 900 meters above sea level in the provinces of Yunnan and Guangdong and Guangxi Zhuang Autonomous Region. Its petals look like tiger's whisker, thus its name. It blooms in April and May. Its slightly poisonous root is used as medicine to treat gastric ulcer, highblood pressure, hepatite and burns.

云南显脉金花茶（Camellia euphlebia var.yunnanensis），山茶科（Theaceae）山茶属（Camellia）。灌木至小乔木，产云南河口海拔360-480米的沟谷雨林中。花金黄色，为名贵观赏花卉，是杂交育种的重要材料。花期2-3月。

Yunnan raised-vein camellia *(Camellia euphlebia var. yunnanensis)* of the *Theaceae* family, *Camellia* genus, can be a shrub or a small tree found in rain forests in gullies 360 to 480 meters above sea level in Nanhekou of Yunnan Province. Its golden attractive flowers open in February and March. The plant is an important material for cross-breeding. It blooms in February and March.

凹脉金花茶(Camellia impressinervis)，山茶科(Theaceae)，山茶属 (Camellia)。灌木至小乔木，特产广西龙州一带，生于海拔500米以下的石灰岩山常绿阔叶林下。花期2-3月。

Concave-vein camellia *(Camellia impressinervis)* of the *Theaceae* family, *Camellia* genus, is a special local variety of the Longzhou area in Guangxi Zhuang Autonomous Region. It grows under evergreen broadleaf trees on limestone mountainsat an elevation of lower than 500 meters. It blooms in February and March.

石笔木(Tuteheria championii)，山茶科(Theaceae)，石笔木属(Tuteheria)。产云南，生于海拔1000-1300米的阔叶林中。花色美丽，花期4-5月。

Stone pencil wood *(Tuteheria championii)* of the *Theaceae* family, *Tuteheria* genus, grows in broadleaf woods 1,000 to 1,300 meters above sea level in Yunnan Province. Its beautiful flowers open in April and May.

锥头麻（Poikilospermum suaveolens），荨麻科（Urticaceae），锥头麻属（Poikilospermum）。产云南，生于海拔600-900米的沟谷密林中，攀援于大树上。花期5-6月。

Awl-tip nettle *(Poikilospermum suaveolens)* of the *Urticaceae* family, *Poikilospermum* genus, grows in gullies in dense forests 600 to 900 meters above sea level in Yunnan Province. It climbs up big trees and blooms in May and June.

缅甸树萝卜 (Agapetes burmarica)，越橘科 (Vacciniaceae)，树萝卜属 (Agapetes)。产云南南部，生于海拔720－1400米石灰山疏林中，附生于树上或岩石上。花期4-5月，钟状花成簇下垂、美丽，供观赏，茎肿大部分供药用。

Burmese cowberry *(Agapetes burmarica)* of the *Vacciniaceae* family, *Agapetes* genus, grows on trees or rocks in sparse woods on limestone mountains 720 to 1,400 meters above sea level in southern Yunnan Province. It blooms in April and May and its bell-shaped flowers hang down in clusters. The swollen stem can be used as medicine.

白花灯笼(Clerodendrum fortunatum),马鞭草科(Verbenaceae),大青属(Clerodendrum)。产广东、广西、海南等地,生于海拔1000米以下的村边、路旁、旷野、荒地或灌丛中,全株入药,味微苦,性凉,有清热解毒、消肿散瘀之效。花期4-5月。

Whiteflower lantern *(Clerodendrum fortunatum)* of the *Verbenaceae* family, *Cerodendrum* genus, grows near villages, by the road, in open wasteland or in bushes lower than 1,000 meters above sea level in the provinces of Guangdong and Hainan and Guangxi Zhuang Autonomous Region. The plant tastes slightly bitter and is used as medicine to expel fever and abate swelling. It blooms in April and May.

赪 桐(Clerodendron japonicum),马鞭草科(Verbenaceae), 赪 桐属(Clerodendron)。产西南、中南等地,生于海拔600-900米的林缘、沟边灌丛中。花期3-4月,盛开时红色一片,煞是美丽。根与花药用,可祛风湿、清肝、明目,治痔疮、疝气、失眠,叶拔毒消肿。

Red glorybower *(Clerodendron japonicum)* of the *Verbenaceae* family, *Clerodendron* genus, grows on the edge of forests or in bushes by ditches 600 to 900 meters above sea level in southwestern and southern-central China. Its red flowers open in March and April in large groups. Its root and flowers are used as medicine to treat arthritis, liver and eye dysfunction, hemorrhoid, hernia and insomnia, and its leaf can be used to abate swelling.

云南石梓（Gmelina arborea），马鞭草科（Verbenaceae），石梓属（Gmelina）。产云南南部、海南，生于海拔460-1300米的干燥疏林中。花期3-4月，花大、美丽而清香，可供观赏，亦可食用，花作染料和香料，还是珍贵用材树种。

Yunnan Malay catchbirdtree *(Gmelina arborea)* of the *Verbenaceae* family, *Gmelina* genus, grows in dry sparse woods 460 to 1,300 meters above sea level in southern Yunnan Province and Hainan Province. Its large fragrant, attractive flowers open in March and April. They are highly ornamental and edible and can be used as a dye and spice. The wood is of high value as timber.

毛石梓（Gmelina villosa），马鞭草科（Verbenaceae），石梓属（Gmelina）。产于中国华南，生于海拔400-700米的林下或灌丛中，供观赏及绿化。花期4-5月。

Bristly bushbean *(Gmelina villosa)* of the *Verbenaceae* family, *Gmelina* genus, grows in woods or bushes 400 to 700 meters above sea level in southern China. It is used as an ornamental plant. The flowering period is in April and May.

节鞭山姜（Alpinia conchigera），姜科（Zingiberaceae），山姜属（Alpinia）。产云南南部，生于海拔700-1100米的林下。花期6-7月，果和根药用。

Joint-whip jinger *(Alpinia conchigera)* of the *Zingiberaceae* family, *Alpinia* genus, grows in woods 700 to 1,100 meters above sea level in southern Yunnan Province. It blooms in June and July. Its fruit and root are used as medicine.

红豆蔻（Alpinia glanga），姜科（Zingiberaceae），山姜属（Alpinia）。产云南、广东、广西、台湾，生于海拔（100-）600-1200米的林下或灌丛中。花期9-10月。果实药用，有祛湿、散寒、消食功效。

Round cardamon *(Alpinia glanga)* of the *Zingiberaceae* family, *Alpinia* genus, grows in woods or bushes 600 to 1,200 meters above sea level in the provinces of Yunnan, Guangdong and Taiwan and Guangxi Zhuang Autonomous Region. It blooms in September and October. Its fruit is used as medicine to expel inner wetness and coldness and help digest.

长柄山姜(Alpinia kwangsiensis)，姜科(Zingiberaceae)，山姜属(Alpinia)。产云南、广西、广东，生于海拔580-680米的沟谷林下阴湿的环境中。花期4月，美丽，供观赏。

Longhandle ginger *(Alpinia kwangsiensis)* of the *Zingiberaceae* family, *Alpinia* genus, grows in wet land under trees in gullies 580 to 680 meters above sea level in the provinces of Yunnan and Guangdong and Guangxi Zhuang Autonomous Region. Its beautiful flowers open in April.

黑果山姜(Alpinia nigra)，姜科(Zingiberaceae)，山姜属(Alpinia)。产云南南部，生于海拔800米左右的沟边及沼泽地上。6-7月开花，供观赏，药用。

Blackfruit ginger *(Alpinia nigra)* of the *Zingiberaceae* family, *Alpinia* genus, grows by ditches or in marshes about 800 meters above sea level in southern Yunnan Province. Its attractive flowers open in June and July and can be used as medicine.

宽唇山姜（Alpinia platychilus），姜科（Zingiberaceae），山姜属（Alpinia）。产云南南部，生于海拔600-1000米的林缘、沟边较为潮湿的环境中，3-4月开花，花大，苞片白色，极美。

Widelip ginger *(Alpinia platychilus)* of the *Zingiberaceae* family, *Alpinia* genus, grows in wet land on the edge of forests or by ditches 600 to 1,000 meters above sea level in southern Yunnan Province. Its large white flowers open in March and April and are very pretty.

九翅豆蔻（Amomum maximum），姜科（Zingiberaceae），豆蔻属（Amomum）。产云南西双版纳，生于海拔500-900米低山林缘和沟边潮湿的环境中，村寨园圃普遍栽培。果食用或药用。花期4-5月。

Nine-Wing Amomum *(Amomum maximum)* of the *Zingiberaceae* family, *Amomum* genus, is found on the edge of forests or in wet land by ditches 500-900 meters above sea level in Xishuangbanna, Yunnan Province. Villagers plant it in gardens as food or medicine. It blooms in April and May.

心叶凹唇姜(Boesenbergia fallax)，姜科(Zingiberaceae)，凹唇姜属(Boesenbergia)。产云南，生于海拔700-1100米的季风常绿林下阴湿处或草地上。花期7-8月。

Heart-leaf concavelip ginger *(Boesenbergia fallax)* of the *Zingiberaceae* family, *Boesenbergia* genus, grows in shady and wet land in evergreen, monsoon forests or weadows 700 to 1,100 meters above sea level in Yunnan Province. It blooms in July and August.

莴笋花(Costus lacerus)，姜科(Zingiberaceae)，闭鞘姜属(Costus)。产云南，生于海拔800-1000米的林缘或灌丛中。花期5-6月。

Lettuce flower *(Costus lacerus)* of the *Zingiberaceae* family, *Costus* genus, grows on the edge of forests or in bushes 800 to 1,000 meters above sea level in Yunnan Province. It blooms in May and June.

光叶闭鞘姜(Costus tonkinensis)，姜科(Zingiberaceae)，闭鞘姜属(Costus)。产云南、广西、海南，生于海拔200-900米密林下的沟边阴湿处。花期8月，花生于根部，根药用。

Canereed spiralflag *(Costus tonkinensis)* of the *Zingiberaceae* family, *Costus* genus, grows in wet and shady places by ditches in dense forests 200 to 900 meters above sea level in the provinces of Yunnan and Hainan and Guangxi Zhuang Autonomous Region. It blooms in August. Its flowers, sprouting from the base of the stem, are used as medicine.

郁金（Curcuma aromatica），姜科（Zingiberaceae）姜黄属(Curcuma)。产中国东南部至西南部,生于海拔360-1900米林缘草坡中。根茎供制中药材“姜黄”,有行气解郁、破瘀、通经的功用,主治胸闷胁痛、胃腹胀痛等症。花期6-7月。

Aromatic turmeric *(Curcuma aromatica)* of the *Zingiberaceae* family, *Curcuma* genus, grows on grass slopes on the edge of forests 360 to 1,900 meters above sea level from southern to southwestern areas in China. It blooms in June and July, and its root, known in traditional Chinese medicine as turmeric, is used to stimulate the circulation of qi, dissolve petechiae and open channels, and treat depression and pain in the chest and stomach. It blooms in June and July.

姜黄（Curcuma longa）,姜科（Zingiberaceae）,姜黄属（Curcuma）。产云南、广东、广西、福建、台湾,生于海拔800-1300米的林下或草地、灌丛中。花期8月,供观赏。亦可药用,能行气破瘀、通经止痛,并可作黄色染料。

Turmeric *(Curcuma longa)* of the *Zingiberaceae* family, *Curcuma* genus, grows under trees, in meadows or bushes 800 to 1,300 meters above sea level in the provinces of Yunnan, Guangdong, Fujian and Taiwan and Guangxi Zhuang Autonomous Region. Its pretty flowers open in August and can be used as medicine to dissolve petechiae, open channels and relieve pain. Yellow dye can be extracted from the flowers too.

舞女花(Globba recemosa)，姜科(Zingiberaceae)，舞花姜属(Globba)。产中国南部至西南部各省区，生于林下阴湿处。花期6-9月，因花型如少女舞姿而得名。

Dancinggirl flower *(Globba recemosa)* of the *Zingiberaceae* family, *Globba* genus, grows in shady,wet land under trees in areas from South to Southwest China. It blooms from June to September. Its pretty flowers look like dancing girls.

红姜花(Hedychium coccineum)，姜科(Zingiberaceae)，姜花属(Hedychium)。产云南南部，生于海拔800-1300米的林缘草地上。花期6-7月，可供观赏。

Red gingerlily *(Hedychium coccineum)* of the *Zingiberaceae* family, *Hedrychium* genus, grows in grassland on the edge of forests 800 to 1,300 meters above sea level in southern Yunnan Province. Its attractive flowers open in June and July.

大叶山奈(Kaemferia glanga)，姜科(Zingiberaceae)，山奈属(Kaemferia)。产云南、广东、广西、台湾，生于海拔800-1100米的山坡草地上。根茎为芳香健胃剂，有散寒、祛湿，辟恶气之功效，亦可作调味香料。花期5-6月。

Largeleaf resurrectionlily *(Kaemferia glanga)* of the *Zingiberaceae* family, *Kaemferia* genus, grows in meadows on mountain slopes 800 to 1,100 meters above sea level in the provinces of Yunnan, Guangdong and Taiwan and Guangxi Zhuang Autonomous Region. It blooms in May and June. Its root can be made into astomachic, and medicine to expel inner cold and wetness It can also be used as spice.

毛姜花 (Hedychium villosum)，姜科(Zingiberaceae)， 姜花属 (Hedychium)。产云南西南部至南部，生于海拔1200-1500米密林中树上或石岩上。花期2-3月。

Haired gingerlily *(Hedychium villosum)* of the *Zingiberaceae* family, *Hedychium* genus, grows on trees or rocks in dense forests 1,200 to 1,500 meters above sea level in southwestern and southern Yunnan Province. It blooms in February and March.

姜花(Hedychium coronarium)，姜科(Zingiberaceae)，姜花属(Hedychium)。产台湾、广东、广西和云南，生于林中。花美丽，芳香，常栽培供观赏。根茎药用，能解表、散风寒，治头痛、风湿痛及跌打损伤等症。花期4-7月。

Gingerlily *(Hedychium coronarium)* of the *Zingiberaceae* family, *Hedychium* genus, grows in forests in the provinces of Taiwan, Guangdong and Yunnan and Guangxi Zhuang Autonomous Region. It is often raised as an ornamental plant for its pretty and fragrant flowers that open from April and July. Its root is used as medicine to treat cold, headache, arthritis and injuries.Its flowering period lasts from April to July.

海南三七(Kaempferia notunda)，姜科(Zingiberaceae)，山奈属(Kaempferia)。产云南、广东、广西、台湾，生于海拔900-1300米的草地和灌丛中。4-5月开花，先花后叶，美丽，芳香，供观赏；根花药用，能治跌打损伤。

Hainan pseudo-ginseng *(Kaemferia notunda)* of the *Zingiberaceae* family, *Kaemferia* genus, grows in grassland or bushes 900 to 1,300 meters above sea level in the provinces of Yunnan, Guangdong and Taiwan and Guangxi Zhuang Autonomous Region. Its leaves come after the fragrant, beautiful flowers bloom. Its root and blossoms are used to treat injuries and broken bones.

长 果 姜 (Siliquamomum tonkinense)，姜 科 (Zingiberaceae)，长 果 姜 属 (Siliquamomum)。该属仅此一种，中国产云南南部，生于海拔500－800米的热带沟谷雨林中。花期4－5月。该种蒴果稍缢缩呈链荚状，为姜科中所仅有者。

Longfruit ginger *(Siliquamomum tonkinense)* of the *Zingiberaceae* family, *Siliquamomum* genus, is found only in tropical rain forests in mountain valleys 500 to 800 meters above sea level in southern Yunnan Province. It blooms in April and May. Its fruit looks like a chain, the only ginger variety that has such shaped fruit.

版纳姜(Zingiber xishuangbannaese)，姜科(Zingiberaceae)，姜属(Zingiber)。产云南西双版纳海拔600-1300米的林下、坡地灌丛中，花苞片红色，极美，花期2-3月。

Xishuangbanna ginger *(Zingiber xishuangbannaese)* of the *Zingiberaceae* family, *Zingiber* genus, grows under trees or in bushes on sloping areas 600 to 1,300 meters above sea level in Xishuangbannan of Yunnan Province. It blooms in February and March and its red petels are very pretty.

脆舌姜(Zingiber fragile)，姜科(Zingiberaceae)，姜属(Zingiber)。产云南西双版纳低山沟谷密林下草丛中。花期8－9月。

Crisptongue ginger *(Zingiber fragile)* of the *Zingiberaceae* family, *Zingiber* genus, grows in grass in dense forests in mountain valleys in Xishuangbanna, Yunnan Province. It blooms in August and September.

毛瓣山姜(Alpinia malaccensis)，姜科(Zingiberaceae)，山姜属(Alpinia)。产西藏东南部和云南西双版纳，生于海拔500－900米山谷常绿阔叶林下阴湿处。花期3－4月。

Haired-petal ginger *(Alpinia malaccensis)* of the *Zingiberaceae* family, *Alpinia* genus, grows in wet areas in evergreen broadleaf forests in mountain valleys 500 to 900 meters above sea level in southeastern Tibet Autonomous Region and Xishuangbanna of Yunnan Province. It blooms in March and April.

艳山姜(Alpinia zerumbet),姜科(Zingiberceae),山姜属(Alpinia)。产我国东南部至西南部,生于山谷林下阴湿处。花艳丽,常栽培于庭园供观赏。根茎和果实供药用,有健脾暖胃、燥湿散寒之功效,治消化不良,呕吐腹泻。花期3-4月。

Bright ginger *(Alpinia zerumbet)* grows in wet spots in forests in mountain valleys from southeast to southwest of China. Its flowers have bright colors. It is often planted in courtyards. Its root and fruit are of medicinal value good for strengthening stomach function and dispel inner wetness and cold, and for treating vomiting and diarrhoea. It blooms in March and April.

圆瓣姜花(Hedychium forrestii)，姜科(Zingiberaceae)，姜花属(Hedychium)。产广西、云南和西藏东南部，生于海拔200-900米的山谷密林或疏林、灌丛中。花期6-7月。

Roundpetal gingerlily *(Hedychium forrestii)* of the *Zingiberaceae* family, *Hedychium* genus, grows in dense forests in mountain valleys, sparse woods or bushes 200 to 900 meters above sea level in Guangxi Zhuang Autonomous Region, Yunnan Province and southeastern Tibet Autonomous Region.It blooms in June and July.

索引 Index

编 后 语 Postscript

独特、神秘的中国热带雨林和生长其间的种类繁多、多姿多彩的热带花卉，常常使对其进行研究和保护的作者们留连忘返，陶醉不已。久而久之，就萌生了编写出版一本融科学性、知识性、趣味性于一体的图册，将栩栩如生的花卉照片奉献给世人，让更多的读者能欣赏到这些大自然造就的精灵，同时也有利于进一步发掘中国热带野生花卉这一宝贵资源。经过作者们几年的野外努力，从大量摄得的照片资料中选择出中国国产原生种58科140属199种热带野生花卉辑成此书。就其数量而言，虽微不足道，然而，它们却是热带野生花卉植物的典型属种，不少种还属于国家保护的珍稀植物，从一个侧面反映了热带雨林的奇伟和绚丽。

书中植物拉丁学名依据“云南种子植物名录”、“中国高等植物图鉴”，为便于查询，书末附有索引。

本书编写过程中，曾得到汪劲武教授以及其它进行植物研究的同仁的赐教和支持，在此一并致谢。

编者

A great variety of colorful plants and flowers grow in the unique and mysterious tropical rain forests in China. Many scientists and environmentalists fascinated by them have been thinking to put out an illustrated book combining scientific research, popular knowledge and common interest so that more people can appreciate the best creation of Great Nature, and at the same time take further actions to learn about this precious resource of tropical plants and flowers. After several years of field studies and careful selection, the authors decided to include in this book 199 species of 140 genera in 58 families. The number is insignificant but they are typical among tropical wild plants and flowers, and many are rare ones under state protection. This book reflects to a certain extent the greatness and dazzling beauty of tropical rain forests in China.

Latin names in the books are from *A Glossary of Spermatophyte in Yunnan Province* and *Illustrated Dictionary of Higher Plants in China*. An index is attached to the end of this book for reference.

We extend our thanks to Professor Wang Jinwu and colleagues in bontanical research who gave us much help during the compilation of this book.

Editors

编　　委：张裕农　曹孟良　王战强　朱用亨　陶国达
尹嘉庆　熊云翔　黄建国　冯志舟　郎楷永
韩晋云　张启泰　杨增宏
主　　编：陶国达　（中国科学院西双版纳热带植物园）
副 主 编：冯志舟　黄建国
摄　　影：陶国达　冯志舟　郎楷永　张启泰　杨增宏
张宪春　韦毅刚　吴德鄰　黄建国　朱宝华
李光照　林洁云　王达明　杨绍增　李洪伟
李渤生　韩晋云　高江云　李建平
责任编辑：王　勇
翻　　译：刘宗仁
装帧设计：战　军　蔡　荣
封面题字：李宪法

Editorial board : Zhang Yunong , Cao Mengliang , Wang Zhanqiang , Zhu Yongheng , Tao Guoda , Yin Jiaqing, Xiong Yunxiang , Huang Jianguo , Feng Zhizhou,Lang Kaiyong, Han Jinyun, Zhang Qitai and Yang Zenghong
Chief editor : Tao Guoda (Xishuangbanna Tropical Garden Academia Sinica)
Deputy chief editors : Feng Zhizhou and Huang Jianguo
Photos by: Tao Guoda, Feng Zhizhou , Lang Kaiyong , Zhang Qitai , Yang Zenghong , Zhang Xianchun , Wei Yigang , Wu Delin, Huang Jianguo, Zhu Baohua, Li Guangzhao, Lin Jieyun, Wang Daming , Yang Shaozeng , Li Hongwei , Li Bosheng , Han Jinyun , Gao Jiangyun and Li Jianping
Managing editor: Wang Yong
Translated by: Liu Zongren
Designed by: Zhan Jun and Cai Rong
Title inscription: Li Xianfa

图书在版编目（CIP）数据

中国热带野生花卉: 中英文对照/陶国达主编; 刘宗仁译. -北京: 中国世界语出版社, 1998
ISBN 7-5052-0395-9

I.中…II.① 陶…② 刘…III.热带-花卉: 野生植物-中国-画册 IV.S68-64

中国版本图书馆CIP数据核字（98）第28701号

中国热带野生花卉

陶国达 冯志舟 黄建国

中国世界语出版社出版

北京博诚印刷厂印刷

（解放军1201工厂）

中国国际图书贸易总公司（国际书店）发行

（北京车公庄西路35号）

北京邮政信箱第399号 邮政编码: 100044

1998年12月(16开)第一版第一次印刷

ISBN 7-5052-0395-9/Q.11

85 - CE - 522S

09800